P9-AOH-910

Good things I wish you

By Manette Ansay

Good Things I Wish You*
Vinegar Hill*
Blue Water*
Read This and Tell Me What It Says
Sister
River Angel
Midnight Champagne*
Limbo: A Memoir

**Available from Headline Review*

MANETTE ANSAY

Good things I wish you

First published in Great Britain in 2009 by
HEADLINE REVIEW
An imprint of HEADLINE PUBLISHING GROUP

1

Cataloguing in Publication Data is available from the British Library

Hardback ISBN 978 0 7553 2990 8
Trade paperback ISBN 978 0 7553 2991 5

Offset by Avon DataSet Ltd, Bidford on Avon, Warwickshire

Printed and bound in Great Britain by Clays Ltd St Ives plc

HEADLINE PUBLISHING GROUP
An Hachette UK Company
338 Euston Road
London NW1 3BH

www.headline.co.uk
www.hachette.co.uk

This book is for W. R.,
to whom I wish all good things.

Contents

Author's Note

This is a work of fiction. Everything in it—from historical sequences to contemporary details—serves, first and foremost, the fictional story I've set out to tell. Those interested in strict historical accuracy should consult the books my characters discuss, debate, and refer to throughout *Good Things I Wish You*. Additional information about the life of Clara Schumann can be found at www.amanetteansay.com.

I wish I could write you as tenderly as I love you and tell you all the good things that I wish you. You are so infinitely dear to me, dearer than I can say . . . If things go on much longer as they are at present, I shall have some time to put you under glass or have you set in gold . . . Your letters are like kisses.

—Johannes Brahms, in a letter to Clara Schumann, 1856*

---

* Berthold Litzmann, ed., *Letters of Clara Schumann and Johannes Brahms, 1853–1896*, vol. 1 (New York: Vienna House, 1973), 73 (hereafter cited as *Letters*).

I wish I could find longing as sweet as you do. It only gives me pain and fills my heart with unspeakable woe.

—Clara Schumann, in a letter to Brahms, 1858[*]

* Ibid., 88.

Part I

The Ax Murderer

The Wine Cellar, 2006

1.

My first date in nineteen years was nearly an hour late. The hostess had brought me two messages, each one saying he was only minutes away, but he was coming from Lauderdale, and even without traffic, that's a long haul to West Palm, where we were meeting in an open-air restaurant. Small tables. Wicker chairs. Below, in a courtyard planted with coconut palms, colorful jets of water rose and fell like expectations. I took another roll from the bread basket, ordered a glass of wine. The dating service, one which demanded lots of money to keep everything off the Internet, had assured me that Hart was "handsome, honest, and caring." Once a week, twice a week, a young woman named America called with yet another recommendation, and all of her recommendations were men who were "handsome, honest, and caring."

"He's an entrepreneur," America had added this time.

"That can mean anything."

"He's forty-seven years old. He has a ten-year-old daughter."

I could tell she was reading from her screen. In the background, other girls just like her—fresh voiced, eager—encouraged other clients.

"He lives too far away," I'd said. "And what kind of name is Heart?"

"H-a-r-t. He enjoys classical music and good conversation. I'm looking at his picture, and he's cute."

"But we'd never see each other."

"If you two kids hit it off," America said brightly, "you'll figure something out."

I was, at the time, forty-two years old; I'd signed up for this service several months earlier, but I'd yet to agree to a date. Too busy, I'd kept telling myself, and this wasn't exactly a lie. There was my job at the university. There was the novel I was supposed to be writing about the nineteenth-century German pianist and composer Clara Schumann and her forty-odd-year relationship with Johannes Brahms. There was my four-year-old daughter, Heidi. There was also the fact that, since my divorce had been finalized, I'd been finding it difficult to make decisions of any kind. Should I put the house on the market? Should I buy green apples or red? Should I find an outside piano teacher for Heidi or keep teaching her myself? The previous week, with the help of my new friend Ellen, I'd finally boxed up the last of Cal's things, odds and ends he'd been promising to collect for months: a framed map of Massachusetts, a shoe box full of pens, an assortment of holiday gifts—candles, boxed jellies, joking plaques—from various junior high students. A swan-necked lamp that had belonged to his mother. Period boots and belts and jackets. Faded T-shirts printed with the dates and locations of Revolutionary War reenactments. Ellen pulled a tomahawk from a dark leather pouch; she wore a man's powdered wig on her head.

"What do you miss about this guy?" she'd said.

"Everything," I'd said. "And nothing."

Now, as the waitress arrived with my wine, I considered what to do with the boxes. Should I mail them to Calvin? Leave them at the graffiti-spattered Goodwill trailer next to the I-95 overpass? Wait until he picked up Heidi for the weekend, insist he take everything along? Each of these options seemed fraught with consequences, all of them unpleasant and inevitable. The box would be lost. I'd be carjacked at gunpoint. Calvin would be angry. The rational part of my brain, the part I recognized, reminded me that I was being ridiculous. But the other part—its nervous newborn twin—was persistent, hungry for disaster. One wrong step, one bad choice, and the worst would happen, the earth would swallow me whole, and if that happened, *when* that happened, what would become of Heidi? Each night, I got up to check windows and doors, making certain that everything was locked. I stayed off the phone during storms. I'd stopped taking vitamins, worried about choking, about Heidi finding me dead on the floor.

By the time Hart showed up, I'd finished my wine as well as the contents of the bread basket. My first impression was that he was utterly exhausted: ashen-faced, pale-lipped, a quietly aging man. I was looking tired myself these days, the bags beneath my eyes worse than usual.

Already you have something in common, said the thin, ironic voice inside my head, and I wished I had left ten minutes earlier, the way I'd wanted to. I should have been at home, tucking Heidi into bed. I should have been reading student manuscripts. I should have been going through the hundreds of pages I'd already written on Clara and Brahms, all of them perfectly fine pages of writing, and not a single one of them right. Not a single one offering fresh insight into the questions others had already asked.*

What was the true nature of their relationship?

Why did the two never marry, even after Robert Schumann's death?

"This will never work," Hart announced, voicing my own thoughts as he sank into a chair. "It is over an hour to get here."

He spoke with a light German accent. Maybe Czech. Too bad I'd never know which. "I told them the distance was a problem," I said, reaching for my purse.

He glanced at me without interest. "You are leaving?"

"My sitter goes home at eight."

"It is seven."

"The traffic."

"Ah."

---

* The definitive book on All Things Clara is *Clara Schumann: The Artist and the Woman* by Nancy Reich. Also recommended: *Johannes Brahms: A Biography* by Jan Swafford. Three recent novelizations, in English, include *Clara* by Janice Galloway, *Longing* by J. D. Landiss, and *Trio* by Boman Desai. As an adolescent, I read and reread *Clara Schumann: A Romantic Biography* by John Burk.

German, I decided. My parents spoke it as children. Of course they stopped when they started school, and then there was the war. Growing up, I'd begged for German words as if they were pieces of hard candy, delicious but unwholesome somehow, certain to rot my teeth.

"I could eat something quick," I said, wavering. Perhaps he might be someone who could help me with translations. "Maybe some soup."

"You like soup?"

"Why not soup?"

He touched the empty bread basket. "You seem to like bread, too."

The waiter nearly tripped in his eagerness to get to our table, and I took a second look at my date: expensive watch, tailored shirt, full head of curly dark hair. This was a man who would always be led to the table marked *Reserved*. I made up my mind to dislike him. The waiter stood ready with his pen.

"I must have more than just soup," Hart said. "I am coming straight from work."

"I also came from work." It seemed important to establish that I, too, had been put out.

"From your university," he said. "America is telling me this. But she wouldn't tell me where. In case I am the ax murderer, I suppose."

I glanced at him sharply. The waiter bobbed and smiled.

"The *se*-ri-al kill-er." Hart landed on all the syllables, striking each one like a clear, hard note.

"We have fresh calamari," the waiter said.

"Of course you do," Hart said.

He ordered soup for both of us, a plate of calamari for himself. Now we were committed. We sat for a moment in silence.

"I *could* be the serial killer," he said, still musing.

"*I* am the serial killer," I said.

For the first time, he looked at me directly. His mouth was small, precise as a comma, even when he smiled.

"That's right," he said happily. "You never can tell. This is such a fucked-up country."

2.

THIS SUDDEN RUSH OF déjà vu: it had happened to me twice before. Isn't it caused by some chemical glitch, a misfiring deep in the brain? It's like becoming aware of gravity, just for a moment, and without warning. It's the same inevitability one feels at the start of a steep, accidental fall.

3.

THE FIRST TIME, I was still in high school. I'd been accepted into the studio of a well-known piano teacher who'd had some success as a concert pianist before rheumatoid arthritis ended his career. This teacher was in his early forties, a soft-spoken man whose handsome face had been damaged by illness and disappointment. Initially, our lessons had been held at his university, but one day he suggested that I come directly to his home. It was closer to where I lived, and besides, he'd have more time for me there.

And that was where it happened. When he opened the door to greet me, I recognized the antiseptic smell of the air, the tint of the linoleum covering the foyer, the blue chintz curtains in the living room windows, even though I'd never been there before. I knew exactly how he'd mispronounce my name (*Jean*-ette instead of Jean-*ette,* he'd say), and that when he did, he would continue to do so for the next four years, making it a little joke between us.

"Go ahead, Jeanie," my mother said—I was fourteen at the time, so Mom still drove me to lessons—and with that, I stepped forward into his house, into his life. What else could I have done? Inside, we passed his very young wife, who nodded and smiled unhappily.

Perhaps she'd already recognized me, too.

Most days, she'd be cleaning when I arrived: mopping floors with Mop & Glo, wiping countertops with Windex, dusting the furniture with lemon-scented Pledge. Erasing the evidence of their lives. There was also a child, a boy with a temper.

He stared at me with shining eyes until I looked away.

Sometimes, during my lessons, which were held in a studio at the back of the house, this boy would start to scream. He could project into every room, even through the thick studio walls, cries that stained the air like smoke, pooled between the pianos' dark legs. I played Chopin, Beethoven, Bartók, Prokofiev. I played Brahms and Scarlatti and Bach. Music was the only light by which I could imagine any future, and by the time I was sixteen, I was coming three times a week to sit at one of the two grand Steinways. Sometimes the piano teacher sat on the opposite bench. Sometimes he sat beside me. I never knew which it would be. He charged ninety dollars for a two-hour lesson, but for me, the cost was twenty dollars a week, what my parents could afford. From time to time, he'd assure them he'd reduce his fees even more if necessary, that he'd always make room for talented young students, that he'd never be accused of turning away the next Clara Schumann because she couldn't give him what he asked for.

It was this teacher who'd first told me Clara's story.

How Clara's mother, Marianne, left Clara's tyrannical father, Friedrick Wieck, after falling in love with another man. How, from that moment on, Wieck claimed Clara as his own, determined to create a child protégée who would become not only a world-class performer (a so-called *reproduction artist*) but a world-class composer, the first woman to join the ranks of Bach, Mozart, Beethoven. How Goethe immortalized her, at nine, by claiming, *She plays with the strength of six boys.* How she'd debuted at the Gewandhaus before she'd turned ten, soloed at the age of eleven, been named Royal and Imperial Chamber Virtuosa by the emperor of Austria at the age of eighteen. Poems were written in honor of her fingers. Cafés served torte à la Wieck, so named for its texture, which was said to be as airy, as light, as the young Fräulein's touch.

My Clara, Friedrich Wieck liked to say, *wasn't raised to waste her life on domestic bliss.**

There were times when the boy's screams became apocalyptic. Eventually, the piano teacher would place a swollen hand on my shoulder, push himself to his feet. There was always a moment when I'd doubt I could bear, this time, the full weight of his rising, but then he'd be walking away from me, unlocking the studio door. This was my cue to stop playing—for all this time I'd have been drilling

* ". . . my father always scoffed at so-called *domestic bliss.* How I pity those who are unfamiliar with it! They are only half alive!"—Clara Schumann, in a diary entry shortly after her marriage. Quoted in Gerd Nauhaus, ed., Peter Ostwald, trans., *The Marriage Diaries of Robert and Clara Schumann* (Boston: Northeastern University Press, 1993), 63.

a cadenza, reworking tricky fingerings—then release my own pent-up breath, rub my shoulders, roll my neck. Perhaps I'd stand, retuck my shirt into my jeans. Perhaps I'd examine the framed prints on the walls: Canaletto's Dresden; the piano teacher, as a young man, competing at the Van Cliburn, portraits of Clara as a heart-faced preteen, as a twenty-year-old celebrity, at the piano with Robert Schumann, whom she married, at last, over her father's objections, warnings, outright threats. Longing for exactly what Friedrich scorned: *domestic bliss.* Longing to escape the same unhappiness her mother, Marianne, had left behind. I thought it was romantic, those years of separation in which she and Robert were forced to see each other only briefly and in secret, corresponding under false names through the help of sympathetic friends.

The screams increased and then abruptly ceased.

In the silence, I thought of my teacher's hands, the fingers splayed, extended, as if he were trying to touch something just out of reach. He'd consulted with a number of surgeons, trying to find someone who'd agree to break his fingers, reset them into the curve he cupped whenever he shaped my touch to the keys.

Protect your hands, he'd say. *Your hands are everything.*

He gave me original compositions to play, lamenting his own inability to perform them. They were layered with long, lyrical passages gliding like oil over hard, dissonant beats. He liked to stand behind me as I played, resting his chin on the top of my head as the three Claras—

budding girl brilliant artist Schumann's fiancée
—watched us, in triptych, from their frames.

I love you like a father, he'd say, *but you won't listen to me.*

4.

BY THE TIME THE waiter returned with more bread, calamari, two steaming bowls of soup, Hart was talking eagerly, bubbling like a pot. His company, it seemed, developed some kind of vision-enhancing technology, but he waved away my questions and told me, instead, about growing up in East Germany. About being selected, at fourteen, to train as a swimmer at a state-run boarding school. About the years he'd spent in the military before finding himself, as an untrained nurse, in charge of a local ER. About medical school in East Berlin. About lying in bed in '89, just after the wall came down, thinking, *I must pinch myself, this cannot possibly be true.* About traveling to Stockholm, Osaka, Miami on a series of research grants. About experiments he'd conducted on Müller cells, a type of glial cell—had I heard of such a thing? Describing these cells, he spoke for ten minutes without apology or self-consciousness. It was not enough for me to say I understood. He batted at my arm: did I *see*? His face was alert now, lit with bright angles. All signs of exhaustion had vanished. America was wrong; he was anything but cute. He was a strikingly handsome man. At the university, I worked each day with writers, scholars, thinkers. But I couldn't recall when I'd last spoken to anyone who had wanted, so deliberately, to teach me something new. Someone who chose his words with such precision, with such passion. With the absolute attention of prayer.

And yet, as I blew on my steaming soup, I couldn't shake the feeling that I'd met him before. Of course, it might just have been the accent. Like my parents, my father in particular, he spoke with the corner of each sentence turning down. Or perhaps we were related somehow. Things like this did happen.

"Where in East Germany did you grow up?" I asked, but even as I did, I suddenly knew the answer.

"In Leipzig. It's—"

"I know Leipzig. It's the birthplace of Clara Schumann."

"Who?"

"Clara Wieck," I said, reverting to Clara's unmarried name on the off chance that he'd heard it. "She married the composer Robert Schumann in 1840. I am writing a book about her lifelong friendship with his young protégé, Johannes Brahms."

"You have been to Leipzig, then? For research?"

"I'm going in July."

Hart gave me an odd, quizzical look. "When in July?"

Below, in the courtyard, the water show started again. Red and green lights traveled up and down our water glasses, flashed inside the bellies of our spoons. Hart would be traveling to Leipzig, too, a few days earlier than I. Of course our visits would overlap. Together we examined our empty soup bowls.

"Funny coincidence," I finally said.

"If you believe in such things. In their significance."

"Actually, I don't."

Hart looked relieved. "I don't believe in them either."

"I *wish* I could believe."

"But you can't." He speared the last calamari.

"Yes and no," I said. "It's hard to explain."

But he'd been talking so sincerely, so intensely, that I felt obliged to try. "Mostly I think it's just a matter of paying attention. Everything is significant, but when you take note of something in a particular way, it winds up changing how you react, how you feel. Maybe just a little, but there it is. Over time, it starts to make a difference."

His expression was impossible to read.

"So if I decide to consider this significant, it could become significant. But I don't believe there is any intrinsic"—I paused, feeling stupid now—"well, significance. Am I making sense?"

Apparently not.

"I am a rational person," he said. "I cannot believe in such things."

I'd been looking into his face; now I looked away. This was something Cal used to say. One cannot exchange ideas with a *rational person* any more than one can argue with a religious fanatic. The night before, I'd sat up rereading the diary Robert began for himself and Clara on the first day of their marriage. Robert's scant observations present his own point of view as the reasonable one, the rational one, and Clara must have believed this was true, for she overwhelmingly supports his assertions, corrections, ideas. Robert was eleven years older, more educated, more entitled—one could argue—to opinions of his own. Yet he could also be blindingly jealous. Childishly petulant. Possessive. He suffered chronic depression, bouts of paranoia, and eventually, auditory hallucinations. He raged and wept and retreated to his piano, where he composed in manic bursts. He recorded compulsively, in his personal diary, details that included how often he and his wife had sexual relations.

And what about those relations? What about that sad, sick man who came to her with reeking breath, unwashed and wild eyed, muttering about angels? The last years of that marriage must have been simply unbearable. Still, she defended him, protected him. She tried to conceal what was happening, even from their closest friends. Even, at first, from Johannes Brahms, who arrived at their door in September 1853, just a few months before Robert, plagued by the voices of spirits, attempted suicide by throwing first his wedding ring and then himself into the Rhine.

Brahms, fifteen years Clara's junior, who would eventually become privy to every family secret.

Brahms, who—according to rumor, then and now—would eventually become Clara's lover.

"No one can be rational about everything," I said to Hart, becoming aware of the silence between us. "Especially when it comes to relationships."

To my surprise, he nodded. "You are speaking of love at first sight, I suppose."

"I don't know. I've never felt anything like that."

"That's because you, too, are a rational person."

I thought of the new, fearful voice in my head. "Oh, is that what I am?"

Again, that sudden smile. "More rational than I. The first time I married, it was love at first sight. At least I thought it was love."

What I thought: "The *first* time?"

What I said: "So was it? Love, I mean?"

"Sure, sure. It could have been love. Why not?"

"Could have been," I repeated.

I felt as if I were on the edge of learning something significant: about life, about love, about my own future. In the courtyard, a band was setting up beside the fountain. In the sky: a hard slice of moon.

Hart took a handkerchief from his pocket, blew his nose.

"Or maybe it was just the hormones," he said. "Who can really say?"

I have such an urgent desire to see you, to press you to my heart, that I am sad—and sick as well. I don't know what is absent in my life, and yet I do know: you are absent. I see you everywhere, you walk up and down with me in my room, you live in my arms and nothing, nothing is real.

—Robert, in a letter to Clara, 1838*

---

* Nancy Reich, *Clara Schumann: The Artist and the Woman* (Ithaca and London: Cornell University Press, 1985), 84.

5.

ROBERT SCHUMANN HAD HIS first nervous breakdown at fifteen. He recovered and went on to law school, but he dreamed of a career as a concert pianist, and, at twenty-one, he became a boarder and pupil at the home of the famed piano instructor Friedrich Wieck. There he encountered eleven-year-old Clara, who'd already mastered techniques that he could only approximate. His own progress, by comparison, at Wieck's *Piano-Fabrik** was slow. When Wieck left with Clara for an eight-month Parisian tour, Robert—in a vain attempt to strengthen his fingers—permanently damaged the nerves in his right hand, with a device of his own invention. To no avail, he tried the remedies of the day: poultices, rest, and *Tierbaden,* the latter of which involved inserting his injured hand into the fresh-slaughtered body of an animal.

It was 1832. It was Leipzig, Germany. The city boasted

---

* Literally translated as "piano factory." Wieck sold pianos, gave lessons, and boarded promising students in the upstairs rooms, where he lived with Clara, her younger brothers, and, eventually, his second wife and their children.

150 bookstores, 50 print shops and 30 newspapers, in addition to the dazzling Fräulein Wieck, who returned, triumphant, from Paris.* By then, Robert had realized that, for him, a concert career was not possible. He turned to composition instead, and soon he was relying on Clara to interpret and perform what he'd written. At first Friedrich looked on with approval. It would be good for the girl to associate herself with this gifted, if controversial, composer. It would inspire her own compositions, already under way. But by 1835, Friedrich couldn't help but notice the way Clara grieved over Robert's engagement to Ernestine Von Fricken, another young piano student living in one of the boarders' rooms above the *Piano-Fabrik*. A hastily scheduled five-month concert tour did nothing to lift his daughter's mood, though she brightened considerably upon her return, at which point in time, we now know from letters, she and Robert became increasingly attached, increasingly committed to each other.

In fact, Robert was still engaged to Ernestine, but Wieck's suspicion on that front was only one of his objections to the match. Schumann, he said, was a drunkard. He was mentally unstable. He'd be unable to provide Clara with the financial and emotional support she needed. She'd sacrifice her artistic gifts to a life spent bearing children, managing a household, buoying up another's talents—

---

* Nancy Reich's excellent chronology (pp. 17–20) in *Clara Schumann: The Artist and the Woman* lists the destinations and durations of Clara's major concert tours. The figures are extraordinary.

No.

Wieck forbade the couple to see each other. He forbade any correspondence between them. When, in 1837, after eighteen months of separation, Robert formally asked Wieck's permission to marry Clara, Wieck's response was to set out with his daughter on a grueling Viennese tour that would last for the next seven months. He pocketed the proceeds of these concerts, leaving Clara without money for necessities. Still, she would not renounce Schumann. In response, Wieck barred her from his house. He held her piano hostage, along with everything else she owned. In the end, she and Robert were forced to take Wieck to court, where at last they were granted permission to marry. This they did on September 12, 1840, the day before Clara's twenty-first birthday.

Clara's mother, Marianne—to whom Clara had turned for refuge—signed the wedding license.

One can't help but imagine that she did so with a vengeance.

And yet, both parents must have grieved to see Wieck's worst predictions come to pass. Clara's first child, Marie, was born after less than a year. Seven more children would follow, including one who did not survive. Whenever Robert composed on his piano, Clara couldn't play her own—he found the noise too distracting—which meant that her practice time was confined to the hour or two, at the end of each day, when he strolled to his pub for a beer.

"My piano playing again falls completely by the wayside," she wrote in the marriage diary, "as is always the case when Robert composes. Not a single little hour can be found for me in the entire day! If only I don't regress too much!"* For a while, she continued to book concert tours, but Robert's growing litany of symptoms—exhaustion, difficulty in speaking, even temporary blindness—prevented him from traveling with her. Left alone, he grew wretched, dangerously depressed.

"This desolation in the house, this emptiness in me!" he wrote in a letter that reached her in Copenhagen. "Letting you go was one of the most foolish things I ever did in my life and it certainly won't happen again. Nothing tastes good or right. On top of that, I really am not well at all—"†

Wieck had been right on that count, too. Robert's mental and physical health continued to deteriorate. The couple moved first to Dresden and then to Düsseldorf, searching for better air, different doctors, any promise of relief.

It was in Düsseldorf, fourteen years after their wedding, that Robert would tell Clara over the breakfast table, *I am not worthy of your love,* before slipping out of the house in his dressing gown and slippers, hurrying toward the Rhine.

The cold February air like a good taste in his throat. Symphonies of angels singing from the rooftops.

Already he was twisting his wedding ring from its tight embrace of his finger.

---

* Nauhaus and Ostwald, *The Marriage Diaries*, 84.
† Reich, *Clara Schumann*, 113.

You can't belong to him and to me at the same time. You will have to leave one, him or me.

—Robert, in a letter to Clara, 1838*

* Ibid., 88.

You are too dear, too lofty for the kind of life your father holds up as a worthy goal; he thinks it will bring true happiness . . . No, my Clara will be a happy wife, a contented, beloved wife.

—Robert, in a letter to Clara, 1838*

---

* Ibid., 86.

6.

Now I was the one who was bubbling away, but Hart didn't seem to mind. The lines between his eyes deepened as he stirred his espresso with a small, bright spoon. I told him about the fishermen who pulled Robert from the river, the medal of valor they received, on display now at the Schumann house in Zwickau.* I told him how, before Robert was brought home, Clara fled to the house of a neighbor, leaving her children in the care of servants, until Robert could be transported, by anonymous carriage, to an asylum outside Bonn. I told him how—ostensibly on the advice of Robert's doctors—Clara would not see her husband again, not even once, until three days before his death. By then, it was July 1856. She hurried home from a concert tour in England, but instead of heading to Robert's bedside—or to see her children in Düsseldorf—she met Brahms for a visit to the Black Sea. From there, they traveled to Bonn, where Robert recognized Clara and tried to

* www.schumannzwickau.de.

embrace her. Two days later, he died, alone, while Clara accompanied Brahms to the train station.*

"It is obvious what happened, isn't it?" Hart said. In the courtyard, the band had started to play; he pulled his chair closer to mine. Around us, the tables had filled with other couples, groups of laughing women, people shouting to be heard. "She fell in love with the younger man."

"Initially, at least, the younger man fell in love with her and Robert, as a couple. No, not like that," I said as Hart raised his brows. "The first time Brahms met the Schumanns, he ended up staying in Düsseldorf for a month. Every day he'd have lunch with them, walk with them, play with the children, all of whom were studying piano. You can imagine the household—it must have been simply ringing with music. And of course Brahms shared all his compositions with Robert and Clara, who critiqued them, championed them, helped him find sympathetic publishers. Every musical genius of the time passed through that house: the Mendelssohns, the violinist Joseph Joachim, Liszt. How romantic their lives must have seemed!"

Hart sipped his wine. "What was the cause of Robert's death?"

"Advanced syphilis. Clara never knew."

"Romantic, indeed."

"He contracted it as a student. Actually, Wieck's resistance to the marriage probably saved Clara's life. By the time she and Robert married, Robert was no longer contagious."

---

* Reich, *Clara Schumann*, 157.

"Just destined for the nuthouse, poor sucker."

"Though Clara insisted he was getting better. And Brahms returned to Hamburg believing this was so. Then, in February, Robert slips out of the house and—"

"Splash," Hart intoned.

"Splash," I repeated. "As soon as Brahms gets word of it, he returns to Düsseldorf to help. This is in 1854. In 1855, Clara moves the children to a new apartment, and Brahms takes a room in the very same building."

Hart laughed. "So it's just as I said. Except the younger man fell in love with the older woman."

"Not necessarily. First of all, Clara was expecting her eighth child when Robert was sent to the asylum. Brahms would have been there for the final months of the pregnancy when she was particularly miserable, uncomfortable—okay, I'll say it, unattractive. And then, as soon as she recovered from childbirth, she took off on a three-month concert tour of northern Germany. At Christmastime she came home for a while, but then she left for Vienna, and after that she went to—oh, Budapest, I think. Then Prague. Finally England. Wherever she could book an engagement."

"She needed the money."

"Of course," I said, "but Robert had been the musical director of the Düsseldorf orchestra, and the city voted to continue his salary, at least for that first year. So her motives couldn't all have been financial, especially since there was still reason to believe that Robert was coming back home."

"Perhaps she felt Brahms to be too much of a temptation."

"More likely she just really missed her first love: performing. As soon as Robert was hospitalized, she avoided both her husband and, frankly, her children, too. Not that this was a completely conscious choice. But after working so hard to restart her career, she must have been terrified that someone—or something—would bring it to a halt yet again. Brahms, at the time, encouraged her work, particularly her composing. While she was on tour, this twenty-something kid took over the management of her household: servants, children, bookkeeping, all the details that were driving her crazy during the years she was Schumann's wife."

"Because this twenty-something kid was in love with her."

"Clara and Robert had helped him with his career. He would have been eager to repay that debt. And of course he had the run of the house, which included access to Robert's library and piano, so there was something in it for him, too. At one point, Brahms's mother writes and says, Look, what you are doing is admirable, but you've got to think of your own career and let Frau Schumann fend for herself. Brahms ignores this. By the way, Frau Brahms was *seventeen* years older than her husband, but there's nothing in her letter to indicate she thought there was anything inappropriate going on."

"*Inappropriate,*" Hart repeated, as if he were tasting the word. "I believe you are meaning *hanky-panky.*"

His diction, coupled with its delivery, cracked me up.

"I doubt there was hanky-panky. Most biographers agree. No hanky-panky, no talk of marriage. Some initial infatuation, I'll grant you that—particularly on Brahms's end—but the relationship was never a physical one. The question, of course, is why."

Hart was shaking his head. "There is *always* hanky-panky."

"You have to consider how Clara was raised. She was very much her father's daughter: proper, concerned with appearances. Also, she was quite naive. She'd fallen in love with Robert as young as fifteen or thirteen, depending on how you interpret things. Even if she'd had feelings for Brahms, she might not have fully understood them."

"Please."

"In her diary, she writes about feeling like Brahms's mother, about taking on that role after his own mother dies. And Brahms had been both beloved and esteemed by Robert. Clara could have been simply continuing a friendship he'd encouraged, even from the asylum."

But Hart wasn't listening, intent on his own thoughts. This was something I could understand. All my life I've been accused of not paying attention when, in fact, I have been listening with that other, inner ear. Couples in the courtyard were dancing now, and I watched them as I waited. The men led the women with such command. The women didn't miss a step. At last Hart said, "She must have been a beautiful woman, this Clara, to interest a twenty-year-old man."

"More like compelling."

"A beautiful woman is always compelling."

"But a compelling woman isn't necessarily beautiful," I said. "I've got a photo of Clara taken in 1854, just after Robert's suicide attempt—"

"You have an actual photograph?"

I am a rational person.

"Okay, a *copy* of a photograph. My point is that, as many times as I've looked at it, I still find it deeply affecting."

The photo shows a thirty-five-year-old woman more beautiful than any girlish image, but untouchable, unreachable, her gaze that of a stone. Her eyes are hollowed by weariness. Her shoulders slump. One gets the sense that, as soon as the exposure is complete, she'll quickly turn back toward whatever darkness lies waiting for her full attention.

"My piano teacher gave it to me," I continued, "for my sixteenth birthday. He said I could learn something by looking at it. He said it would help me understand things about men and women most people don't figure out until after it's too late."

"Things," Hart said, "such as those you are writing about in this book?"

This pleased me. "Yes," I said.

"You had a relationship with this teacher?"

There was no segue between topics. There was no change in the tone of Hart's voice, in the close set of his mouth. He was studying me coolly, impersonally.

"We were friends, of course," I said, trying to sound nonchalant, "if that's what you mean by relationship."

"An *inappropriate* relationship?"

I pictured Hart in a tile-floored research lab: white coated, peering into a microscope. "Alas, no. I was a nice Catholic girl."

"I do not believe you," he said.

"He was thirty years older than me," I said, hating the pleading note that had crept into my voice. "And married. And, anyway, he's dead now. Someone sent my mother the obituary."

"What I meant," Hart said, "is that I do not believe you were ever a nice Catholic girl."

"Oh." I could feel myself flush.

"Besides," he said, signaling the waiter, "men and women can never be friends."

This, then, was the end of our date. Hart paid the bill, waving my hand away, and then, as if released from chains, we stood up at exactly the same time. To my surprise, he gave me his business card, so I handed him mine.

"Maybe you wouldn't mind helping me," I said, "if I get stuck on a translation."

"Sure, sure. We will talk on the phone. It's the way these things go. Do you want to go flying with me some time?"

"Flying?" I said, confused. "Like, in a plane?"

"No-no-no, not a plane," he said. "A glider."

"I don't even know what that is," I said, and then, as I was reading the name of his company (Viso-Tech) and his full name (Reinhardt Hempel), he asked how my marriage ended. Again, there was no transition between topics. When I didn't answer, he answered the question for me.

"I suppose he was unfaithful," he said. "It is the usual thing."

"Calvin wasn't cheating," I said.

"Then you," Hart said, and now he was impatient. "Come on. It is always the same."

Abruptly I awakened from whatever was holding me here. This man was a stranger. I never had to see him again. I could drive home to the safety of my daughter, listen to the music of her breathing as she slept.

"I wish I *had* cheated," I told him. "I wish it with all my heart."

I left him beside the table. The entrepreneur. The rational person. The ax murderer. I did not look back. By now I was more than two hours late; it was my babysitter who would kill me. And indeed, she was waiting for me at the door: twenty-one years old, radiant, furious, her purse already slung over her shoulder.

"Where *were* you?" she said, as if she were the mother and I the recalcitrant child. "Why didn't you call?" And, like that child, I could offer no explanation. All I could do was tell her I was sorry for staying out too long.

*Clara Schumann at Thirty-five**

* Robert Schumann Haus, Zwickau.

Part II

¤

Virtue

7.

HEIDI WAS ASLEEP, TANGLED in blankets, some large and square, some rectangular, one no bigger than a handkerchief. Each had a cool, silky border she liked to scratch as she fell asleep. I rubbed the smallest one against her cheek; she sighed, curled her hand into mine. She'd inherited my wide palms as well as my good ear. Already she was finishing the first Suzuki Book; she could sight-read simple melodies, if she wanted. Lessons were battles, of course, at her age, but hadn't my own been the same? Briefly, I lay down beside her, but I couldn't get comfortable, couldn't find enough space for my adult limbs between her stuffed animals and decorative pillows, puffs of pink and white sheets. Something had unsettled me. Hart had unsettled me. The more I thought about the end of our conversation, the more it made me mad.

Men and women can never be friends.

Outside Heidi's window, coconut palms rattled their thick, dry fronds. Beyond them, in the lake, frogs called to one another—you could hear everything through the glass—and now and then, there came another kind of cry, high pitched and sustained. Nothing close to wilderness is left in Palm Beach County, but this little man-made lake, at least, still offered a bit of peace. Tarpon churned the brackish water. Anhingas spread-eagled in the cypress trees. Great horned owls nested in the strip of woods that divided us from the highway.

Once I would have said that Cal and I were friends. Would always be friends. No matter what.

I slipped out of Heidi's bed, paced through the kitchen and family room, circling twice through the pocket doors before heading down the hall toward my study. Marks from the furniture Calvin had taken with him—the dining room set, the antique chairs, the guest bed—were still embedded in the carpeting, like the fossilized prints from some prehistoric animal. Above my writing desk there's a portrait of a skull, haloed in gold leaf, on a striking blue background. It is one of the most beautiful things I've ever seen, and I bought it directly from the artist, believing I'd tuck it away to look at someday, a comfort, perhaps, as I lay dying. But shortly after Cal moved out, I hung the portrait over my desk. It helped—it still helps—looking up at that portrait. Assuming the old attitude of reverence. Remembering such loveliness exists, regardless of how empty I might feel.

Robert, dying in the asylum, asked for a portrait of Brahms, and while some claim the request was rooted in jealousy, Jan Swafford's biography suggests that Robert merely longed to look upon something beautiful. A suggestion that seems utterly believable to me, especially when you read Robert's letter to Brahms on the subject: *I received your picture through my wonderful wife, your likeness that I remember so well, and I know its place very well in my room, very well—under the mirror.** When you note that Robert welcomed Brahms as a visitor. And how carefully, tenderly, Brahms wrote to Clara of his impressions, putting a bright face on a darkening picture, expressing his affection for them both.

"Do not heed those small and envious souls," Clara wrote in the diary intended for her children, "who make light of my love and friendship, trying to bring up for question our beautiful relationship which they do not understand nor ever could."†

In August 1856, two weeks after Robert's funeral, she and Brahms vacationed together in the Swiss town of Gersau. There they hiked the slopes of the Rigi, boated on Lake Lucerne. Free at last from Clara's marriage, in that beautiful place far from home, something happened between them. Correspondence from that period has been destroyed, some of it at the urging of Brahms, some of it at the request of Clara's oldest daughter, Marie. All we know for certain is that they returned to Düsseldorf separately and sadly. Brahms departed thereafter for Hamburg, while Clara embarked on yet another punishing concert

---

* Berthold Litzmann, ed. *Clara Schumann–Johannes Brahms: Briefe,* vol. 1 (Leipzig: Breitkopf und Hartel, 1927), 188 (hereafter cited as *Briefe;* original translations by Winfried Reichelt).

† John Burk, *Clara Schumann: A Romantic Biography* (New York: Random House, 1940), 324.

tour. Still, they remained—by their own definition—"best friends" until Clara's death in 1896, when Brahms would exclaim, "Apart from Frau Schumann, I am not attached to anybody with my whole heart."*

He died in 1897—jaundiced, obese, alone—eleven months after her passing.

There is no evidence that marriage was ever discussed, either with each other or with mutual friends, to whom each wrote long and intimate letters, most of which have survived.

There is no evidence of hanky-panky.

I stared up at the portrait. The portrait stared down on me.

What happened between them in Gersau? There was something I was missing. The same thing other writers—biographers, memoirists, contemporary novelists—had also struggled to see. Hart's business card floated in my purse; I considered giving him a call. Our conversation didn't seem finished somehow.

But by now it was almost midnight. And it wasn't Hart I wanted to talk to.

Truth be told, I was longing to talk to Cal.

---

* Jan Swafford, *Johannes Brahms: A Biography* (New York: Vintage Books, 1999), 611.

8.

We were married for twelve years, Cal and I. His depression had always been part of our lives. First there were good years, with bad stretches. Then there were bad years, with good stretches. And then there were the years when it settled in for good, never talked of but always present, a phantom presiding over our table, slipping a cold hand into our bed. Cal ate compulsively, but without satisfaction. Evenings he drank, quietly and alone. These were the years he began changing jobs—from private schools to public to private again—looking for the right faculty, the right students, the right administration.

He was restless. He was bored. Increasingly, he was angry.

At first I thought the reenactment trips were a good thing. They gave him an outlet, a focus, a purpose. But each trip served only served to increase his dissatisfaction. Night after night, he'd stay up too late, e-mailing and blogging with people he'd met. I'd hear him on the phone, ranting against colleagues and coworkers, family members, friends. Everyone around him was holding him back. Everyone was secretly against him.

These rants were never against me.

Then, one day, they were against me.

Still, I defended him, protected him. I tried to conceal what was happening, even from the closest of our friends. I waited for the man I'd loved to resurface, the way—I am certain of this—Clara waited for Robert, year after year. And, like Robert, Cal always did reappear, as exhausted as a swimmer who has nearly drowned, though each time his return was less complete. The difference in him was so gradual that at first I didn't understand the change. I thought it was just that we were aging, fading. I thought it was simply a trick of the light.

I am a rational person, Cal liked to say, but there's more to all this than reason. Had we behaved like rational people, there would have been no Heidi, who is the antithesis of reason: filled with emotion, bursting with light, passionate about everything, everything, everything from the fit of her socks to the stroke of a pen.

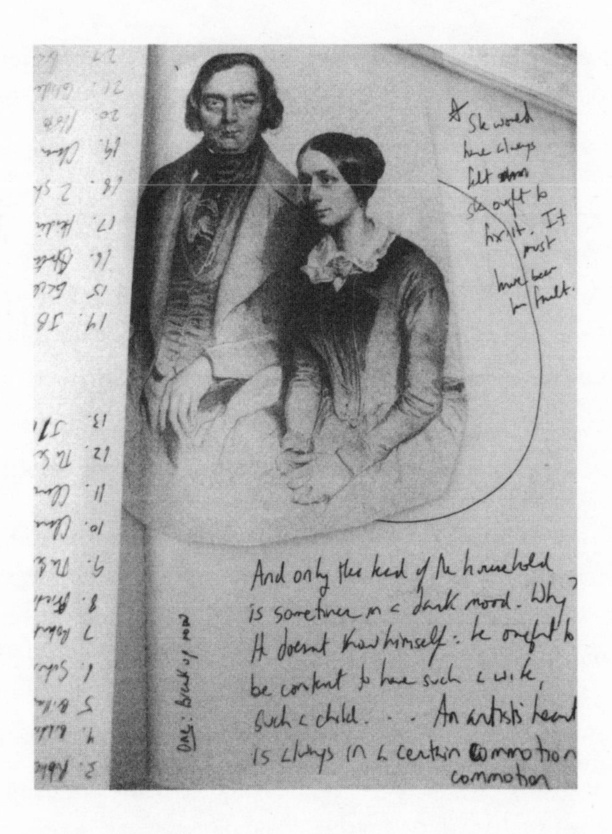

It often worries me that I frequently inhibit Clara in her practicing, since she does not wish to disturb me while composing . . . she sometimes lacks time nowadays, and I am responsible for that and yet cannot change it. But Clara does recognize that I have to nurture a talent, and that I am full of the most beautiful energy right now and still do have to make use of my youth . . . we are really most fortunate to have each other, and understand each other, understand so well, and love each other with our whole heart.

—Robert, in the marriage diary, 1842*

---

* Nauhaus and Ostwald, *The Marriage Diaries*, 178.

Clara has written a series of smaller pieces, more delicate and richly musical in their invention than she's ever achieved before. But having children and a husband who constantly improvises does not fit together with composing. She lacks the ongoing exercise, and that often bothers me, because many a heart-felt thought thus gets lost that she does not manage to execute. Clara herself knows her primary occupation to be a mother, however, so that I believe she is happy under these circumstances, which just simply cannot be changed.

—Robert, in the marriage diary, 1843*

Clara is now putting her Lieder and many piano compositions in order. She always wants to make progress, but on the right Marie hangs onto her dress, Elise also makes work, and the husband sits absorbed . . .

—Robert, in the marriage diary, 1843†

---

* Ibid., 185.
† Ibid., 199.

9.

DESPITE THE URGINGS OF Joseph Joachim, Brahms hadn't wanted to call on the Schumanns. He'd once sent Robert a packet of compositions, but these had been returned, unopened. That was enough for him. As for Frau Schumann, there was no discussion: she was a world-class pianist, a national treasure. Robert Schumann's wife. But a woman, nonetheless, and therefore of little interest. Certainly no different from any other female walking the streets in her long, chaste gown, as if she believed she were fooling anybody. As if she weren't, at heart, the same filth he paid for as easily, as comfortably, as a cool draft of beer on a warm summer's night and forgot, just as easily, afterward.

Sometimes there were those he didn't have to pay, those who admired his beauty.

These were the ones he did not forget.

These were the ones he hated.

It was 1854. He was nineteen years old. Schumann's house might have been the house of anybody, a clerk, a teacher, a banker. Brahms rattled the knocker, then flushed, a letter of introduction clutched in his hand.

Why should I have to come like a beggar? he thought as a little girl opened the door.

His temper evaporated instantly beneath the flat, numb glaze of her stare. Why the plum circles beneath her eyes? Why the darkened parlor like a shroud about her narrow frame? No sign of a maid, a nurse. Something wasn't right. He knew this without setting foot inside the house. Of course, he'd heard the rumors, like everyone else: Schumann mumbling to himself at rehearsals, Schumann catatonic at a dinner party, doctors advising rest and a change of air.

"Johannes Brahms to see Herr Schumann," he told the child, speaking with a gentleness that surprised them both, for she blinked as if he'd only just materialized. As if, until that moment, she'd been looking at empty air.

"Papa and Mama are out walking," she said. She had a pleasing voice—did she sing? "Come back tomorrow at ten."

The door closed in his face without ceremony. Now he'd have to spend an extra night at the cheap guesthouse he couldn't afford in the first place. Enough. He'd pack up his rucksack immediately. He'd take the next train to Leipzig, where, alone, he'd present his compositions to publishers recommended by Joachim, by other powerful friends.

Walking fast through the Düsseldorf streets, he turned his face from the passing glances of hollow-eyed women, from the hard, wet stares of certain men. The trick was to keep moving, to avoid meeting anyone's gaze. Suck in his teeth and narrow his eyes. Knot the pale prettiness out of his skin. At fourteen, in the *Animierlokale**—where his father first found work for him—prostitutes had tugged his hands from the piano keys, lifted him like an infant, passed him from lap to lap. The men urged them on, and there'd been those who'd press hard against him, yes, but that, somehow, had been less disturbing than the women: hags bending over him with painted faces and flickering tongues, choking on the sort of laughter which has little to do with pleasure.

Which has nothing to do with joy.

He arrived at the guesthouse in his darkest mood, and yet he did not pack up his rucksack. Perhaps he was simply too tired to continue on to Leipzig so late in the day. Or maybe he thought more of Schumann's opinion that he'd ever admitted to anyone. He might have been thinking of the child at the door. He might have been thinking of nothing at all. Who can say why we make the choices that come to be seen as significant, ordained?

The next day, he returned to the Schumanns' house at exactly ten A.M. This time it was Robert who answered the door. It was Robert who shuffled, in slippers, to his own Graf piano, which he invited the young man to play. It was Robert who interrupted the C major Piano Sonata after listening for only a minute or two. His voice trembling with excitement. His eager eyes brightening with tears.

"Excuse me," he said, already rising, preparing to speak the words that would change everything for them both, "but I must call my wife."

---

* Waterfront bars frequented by sailors.

Düsseldorf, November 5th, 1853

Dear Sir!

Your son Johannes has become very dear to us; his musical genius has provided us with happy hours. To ease his first steps into the world, I have spoken what I think of him in public. . . . and think that this will give a father's heart a little pleasure.

You may live with confidence towards the future of this beloved of the muses and always be assured of my deeply felt concern for his happiness.

Sincerely yours,

R. Schumann

—Robert Schumann to Johann Jacob Brahms
(father of Johannes Brahms) in Hamburg*

* Litzmann, *Briefe*, 1.

I often have to restrain myself forcibly from just quietly putting my arm around her and even—I don't know, it seems to me so natural that she could not misunderstand.

I think I can no longer love an unmarried girl—at least, I have quite forgotten about them. They but promise heaven while Clara shows it revealed to us.

—Brahms, in a letter to Joseph Joachim, 1854*

* Artur Holde, "Suppressed Passages in the Brahms–Joachim Correspondence Published for the First Time," *The Musical Quarterly* 45, no. 3 (1959): 314.

10.

For the first weeks of 1854, Brahms and the Schumanns fell into happy routine. Mornings, they walked out together. Lunchtime was spent with the children. Afternoons were for the piano, for guests, for preparing Brahms's scores for publication. For listening—often with averted eyes—to Robert's latest efforts, which had been (so Robert claimed) dictated to him by friendly spirits. Brahms watched as Clara distracted anyone present with follow-up performances of Robert's most ambitious works, pieces to which general audiences were still intractably resistant: his fine *Carnaval,* his *Symphonic Études,* his *Kreisleriana.* He saw how, in company, she intercepted any questions directed toward her husband, answering them exactly as Robert might have done, and he began to understand why Robert looked to her, leaned on her.

He would not have believed a woman could be so capable, so strong.

Did I not believe too, before I knew you, he would write to her in 1854, *such human beings and such a* *marriage could only exist in the imagination of the most beautiful minds?*[*]

She was not, by any means, a pretty woman, yet he found himself drawn to her side. He looked for little ways to be of help. He thought of clever, cheering things to say. He sat at the piano when she grew tired, played every request she made. Alone in the guesthouse, he stayed up too late: composing, drinking wine, considering her sweet, weary eyes. Sometimes he imagined her in Schumann's bed—but no. He pushed the thought away. He was not, after all, attracted to her, an older woman, a mother, as plainspoken as she was plain-faced. And her affection for him was straightforward, clear. Blunt. Like her remarks on his manuscripts. Like her pieces of good advice.

He bought a new coat to please her. He changed his fingerings.

Still, he did not suspect her full powers.

Then, one day, as they were finishing lunch, Robert began tapping the tabletop, and though Clara stood up and spoke his name, though baby Eugenie began to cry, though the cook appeared from the kitchen with anxious offers of soup, more soup, *Möchte Herr Schumann ein bisschen mehr?* he pounded away, beating hard time with the flat of his hand as Elise, Ludwig, and Ferdinand pounded, too; as Marie took Eugenie into her arms; as Julie rose coolly from her place and, with a beautiful woman's look of disdain, slipped quietly from the room.

---

* Litzmann, *Briefe,* 10.

"Do you hear? Do you hear the heavenly choir?" Robert shouted as Clara rounded the table, catching her skirt, spilling the water that remained in the pitcher as she stilled, at last, her husband's hand with the wide, warm weight of her own. Silence as Robert lifted that hand to his mouth, kissed it, rubbed his sweating cheek against it, all the while looking into Clara's face as if trying to recall her name. Brahms stared, helpless, along with the children. It was like watching a drowning man pulled from the water, the precarious moment when it seems both the man and his rescuer will tumble into the waves. Then, impossibly, through the sheer force of will, the balance shifts.

She had raised him up again.

Panting, weeping, wild-eyed and wet.

It was the most beautiful thing Johannes Brahms had ever seen.

Today Schumann spoke about a peculiar phenomenon that he has noticed for several days now. It is the inner hearing of beautiful music in the form of entire works! The timbre sounds like wind music heard from afar, and is distinguished by the most glorious harmonies. . . . He spoke of it, saying, "This is what it must be like in another life when we have shed our corporeal selves."

—From the diaries of Rupert Becker,
concertmaster of the Düsseldorf orchestra, 1854*

* Reich, *Clara Schumann,* 142.

11.

One warm evening, in the middle of May, my mother dropped by just as I was finishing Heidi's lesson. She'd brought the piano teacher's obituary, something she'd been promising to show me.

"He was good to you," she said, lifting Heidi off the bench and into her arms for a kiss. "He was a friend."

"We have five more minutes on our lesson," I said. "Heidi, do you want to play something for Grandma? How about Musette?"

"I don't want to play anymore." From the safety of my mother's arms, Heidi shot me a look of triumph.

"He gave you that portrait of Clara Schumann. Do you still have it?"

"I don't know. Heidi, you owe me those last five minutes."

"Remember that time he told me I should take you shopping for clothes? He even gave me some coupons."

"I didn't want to use them," I said. "I didn't want to go."

"Still," my mother said, placing the obituary on top of

the piano, "he meant well. It was a nice gesture, don't you think?"

Heidi asked, "Is that man dead?"

What I said to Heidi: "He had a good, long life."

What I said to my mother: "He was kind of an odd person, if you want to know the truth."

"He certainly loved his students."

"Grandma, will you play a game with me? Will you pretend we are kittens?"

"Fine with me," I said, giving up, and as the two of them decided on the location of a safe, warm nest, I bent over the single smudged paragraph of accomplishments, the face that might have belonged to any tremulous old man.

He wore the same dark wool sweater at each lesson.

He wore slippers with argyle socks, plaid trousers, a wide leather belt.

I wore oversize shirts, baggy jeans, tennis shoes. I kept my arms close to my sides as I slipped between the rows of chairs at the monthly open studios, at the master classes with de Larrocha, Watts, Gutierrez, squeezing myself along balcony edges during group excursions to winter symphonies, to summer festivals, to private concerts where the piano teacher guided me forward, introduced me, invited me to play. Afterward, he'd speak to me sternly about what he called *the problem of your confidence*. It was important, he'd say, to look people in the eye, to speak clearly, to accept a compliment with a smile. As a performer, I'd have to get used to interacting with strangers. I'd have to get used to male attention. And I'd have to learn to dress—was there someone who could help me? My mother, perhaps, or a sister? Would I be offended if he, himself, made a few suggestions?

He was speaking to me as my teacher, of course, but also as a friend.

"A little bit of lipstick never hurt a girl," he said. "Not that I'm suggesting you should paint yourself like one of Herr Brahms's whores."

He apologized for his language. He only wanted what was best. He was trying to protect me, the way Clara's father tried to protect her, but did Clara listen? Did she? Could I understand how much Clara's father had loved her, how he would have done anything to save her from the life she lived with Schumann, reduced to a common hausfrau, wiping the mouths and asses of brats? I must find a man who would be prepared to offer everything, but only from a distance. A man who would look out for me. Who would ask almost nothing in return.

Kissing my fingers, untucking my shirt.

How can you play if you can't lift your arms?

We were at it again, two hands, four. The room echoed with our longing. Still, I wouldn't let him hold me. In the end, I'd always get up, step away.

For art is about desire, is it not, and never its consummation?

"I am beginning to think," the piano teacher said, "that you are incapable of passion."

12.

For several years, there was a man who was in love with me. L— had gotten divorced after discovering a cache of e-mails from his wife's lover, and he'd call me (not too often, of course, for I was married to Cal) just to see how I was doing, talk about books, exchange manuscripts-in-progress. We'd talk about love and relationships, marriage and friendship, true friendship between women and men, which we both agreed was absolutely possible, why not? Not only possible. Necessary. At a literary conference, drunk on wine and success, I told him about Cal, our separate rooms, our separate lives. L— followed me up to my hotel room . . .

. . . where I left him standing outside the door.

All night long, my chest and belly ached with what I thought was virtue.

"I don't see how you can stand it," he said the next time we spoke.

The last time.

"It's not so bad," I said.

"Then something's wrong with you," he said, and I hung up on him.

Shortly after my divorce, I heard from him again, a brief e-mail in which he said he'd been sorry to hear about Cal and me. He'd recently remarried, someone we both knew, a woman who writes about horses. She, too, he said, was thinking of me, would keep me in her thoughts. Both of them had been through it themselves. Both of them knew how tough it could be. So how was I doing? Probably okay. But I shouldn't hesitate to drop him a line if I ever needed a listening ear.

13.

"GOING THROUGH A DIVORCE," Ellen had said as we'd carried Cal's boxes out to the garage, "is like going through chemotherapy. You have to expect to get really, really sick. The difference with divorce is that you know you are going to get better."

*Self-Portrait: Gaela Erwin**

* www.gaelaerwin.com.

Part III

Frozen

14.

THE SECOND TIME I experienced déjà vu, Heidi was two years old. It was October, a Thursday morning. I'd just started at the university, teaching classes on Tuesdays, Wednesdays, and Thursdays. Each Tuesday, I made the six-hour round-trip commute from West Palm Beach, arriving home just in time to tuck Heidi in for the night. On Wednesday nights, however, I'd stay over in Miami. I'd check into my hotel room by six P.M. I'd write until one or two in the morning, sleep in until eight the next day, when I'd start to prep my graduate seminar.

This was the longest block of writing time I'd had since Heidi's birth.

That morning, just as I'd been packing up my computer, I'd received a call from my mother. Cal wasn't working at the time. He'd resigned a full-time contract on the promise of another job that, in the end, hadn't worked out, but we were trying to look on the bright side. He'd been thinking about getting out of teaching anyway. He was thinking about earning a Ph.D. He was supposed to be researching programs, studying for the GRE, but the reality was that he was stuck, not certain what he wanted to do next. If we could have talked about it—but we couldn't talk about it. The trick to getting along seemed to be avoiding talk altogether. I worked and wrote and took care of Heidi. He read the news on the Internet, blogged with reenactment friends.

Of course he took care of Heidi on the days I was at work.

"You don't give Cal a chance with that baby," friends and family members said. "Get out of the house. Get out of his way. Let the two of them get to know each other."

I believed them because I wanted to believe them.

I believed them because I waited all week for that single night, alone at my computer, in room 342 at the Holiday Inn.

"I don't want to worry you," my mother said. "But yesterday, Cal and Heidi came over, and Heidi threw up on her shirt. I offered to wash it, but he said no, he'd take care of it when they got home."

"Is she sick?" I said.

"I think she's fine," my mother said. "But this morning, when I stopped by to see if she was feeling better, she was still wearing the same shirt."

"Maybe it's just that he couldn't get the stain—"

"She stank to high heaven, Jeanie," my mother said. "And Calvin looks—well, he hasn't changed his clothes since yesterday, either. It's more than a hangover, Jeanie. He's not in a good state of mind."

What I should have said: "I'll be home on the next train."

What I said: "I've got to teach in an hour."

My mother did not say anything.

"You can't just cancel a graduate-level class."

My mother said, "You know what you can and cannot do."

Now, nursing a Starbucks cappuccino, I was standing at the corner of Ponce and Dixie, preparing to dash across the four-lane highway, as I did every day, to get to campus. Hurricanes had knocked out the pedestrian walk signs, so you had to time the lights, gauge the speed of the oncoming traffic, which tended to accelerate—Miami being Miami—at the sight of a human being actually braving a crosswalk. During the first week of classes, at exactly this spot, a student had been hit by a car and killed. I thought of her today, as I always did, reminding myself to be careful.

High overhead, a half-dozen vultures circled aimlessly.

Perhaps Heidi had insisted on that particular shirt and Cal simply hadn't had the energy to deal with it. Perhaps she was sick, had been sick all night long, and that's why Cal was in the same clothes: he'd spent a sleepless night caring for her. None of this rang true, but I didn't want to think about what it meant. By *what it meant,* I did not mean what it meant for Heidi or, for that matter, Cal. I was thinking about my writing. I was thinking about myself. I was thinking about what it would mean to lose those Wednesday nights, and I thought to myself, in exactly these words: "I am a dead woman. I am dead."

The light changed. There was a gap in the traffic. I decided to run for the median, wait for a second opportunity to cross the remaining lanes. But as soon as I stepped up onto the narrow concrete strip, I could see exactly what would happen next. Even before I dropped my cappuccino, I knew how it would bounce, still capped, into the oncoming traffic. Even before my left ankle buckled, I knew how my body would feel as it fell, the impact of my hip, vibrations rising through the concrete. The car that would strike me already gaining speed. The steely flash of fender in the split second before, in an attempt to twist away, I'd make things worse by rolling directly into the path of the wheel.

The traffic lights changed. Changed again. I stood, frozen, on the median. How long was I waiting there before a group of students came along? One of them doubled back.

Are you okay? You want to cross with us?

As long as I remained there, the future could not reach me.

Do you want to take my arm?

Even now, I can re-create the feeling of that car striking me as if it really happened, though in fact it never did.

SOUTH
DIXIE HIGHWAY

15.

The night before the suicide attempt was the first time Clara hadn't shared his bed. Sheets wet with sour perspiration. Bruises, the size and shape of kisses, dotting her upper arms. Another bruise on the small of her back from the unexpected bucking of his knee. That was the thing that did it. Not that she would ever blame Robert for what he'd done. Hour after hour, the dead were tormenting him, singing the same three notes while he thrashed like a man in physical pain. At last he'd lulled into stillness. She, too, slept for a while—until awakened by the impact of the floor.

"Stay away from me," he was shouting. "I will hurt you! I will kill you! How can you bear it?"

And she looked up at his frenzied face peering down at what he'd done—fat-cheeked, demonic—and answered, "I cannot."

"Why not fetch your maid to sit with me?" he'd said, suddenly cogent, calm. So she'd kissed his damp hand and done as he had said, crawling into bed beside Marie, where she slept six uninterrupted hours before awakening to the sound of the milk cart. Marie was already gone. A faint, greasy light filtered in beneath the door, seeped through the rift in the curtains. Downstairs, the cook battled the coal stove. The clop of the milk cart continued on. Everything normal. Everything calm. For a moment, a thought came into her mind, accompanied by guilt as sharp as pleasure. But then she heard him talking, heard the maid's stout reply.

The door opened. Marie, fully dressed.

"Papa is better, I think," she said.

Time, once again, to get out of bed. To step forward into the day. Robert already seated at table, fumbling his napkin beneath his great, soft chin. Clara directs the children toward eggs set in cups, toward bowls filled with porridge and cream. She eats. Smiles. Performs. As she's been performing since the age of nine, through broken pianos and cholera epidemics, through her father's rages and her husband's jealousies, through cities and countries and continents not nearly as far-flung as she'd hoped. Even in good health Robert requires just this: familiarity, routine. A house in which children sit in their places. Orderly servants. Clean windows and halls. A room of his own where he keeps his piano, books, compositions, poetry, reviews. Glasses of sweet, dark beer at the pub during the hour or two before supper each night when she closes the door against children and staff, practicing hastily, hungrily.

Later he'll wrap his arms around her with the same quick, anxious greed.

So she's come to be pregnant with his eighth child. And to anyone here in the twenty-first century who might object to the phrase "his child," who suggests that these children belong to them both, she'll insist—urgently, fiercely—that the children, like everything else, belong only to her husband. As the concerts she has performed, the money she has earned, the lessons she has taught belong to him. The few compositions she's managed during her fourteen years of marriage? These belong to him as well. If she could, she'd drain the very blood from her body. She'd feed it to him as she feeds broth to the little ones, trembling at the lip of the spoon.

The only thing she has kept for herself: her desire to play music in public, to perform. Since it's all that she has kept, it's the one thing he wants. Day after day, it sits between them. The single thing she's held back.

But today, as he smiles across from her, the children encircling them both with bright laughter, it seems possible that everything she's offered, everything she's given, will suffice. Perhaps Marie is right this time. Perhaps Papa is truly better. Clara is looking directly at him—a look in which everything good must shine—when he says, "Forget about me, Clara. I am not worthy of your love."

The words hit her like a slap.

Forget about me. Forget about us. Forget about all we have meant to each other.

He said that, she will write in her diary, *he to whom I had always looked up with the greatest, deepest reverence.**

Here it is then: the moment it happens. One small thing is said or done—not the first thing, not the worst—but it cuts, at last, the weathered string that's bound him to your

---

* Reich, *Clara Schumann,* 143.

heart. After this, everything happens in third person. All you can do is watch your own life unfold as if it is happening to a stranger. Even if he hadn't thrown himself into the Rhine, Clara would have left him that day. She'd have stayed with the neighbor lady until arrangements could be made. She'd have wept and wept and wept and, still, she'd have somehow found it in herself to weep more.

Weeks later, at the mention of a visit to Endenich, she dissolves into such hysteria that it's finally suggested, by the doctors, that she send Herr Brahms instead.

Your treasured husband has not changed at all, Brahms writes to her immediately after the visit. *He only became a little stronger. His appearance is happy and bright, his movements are just the same as in the past, one hand held to his mouth. He smoked in short puffs as usual. His gait and his greeting were more free, firmer, which is only natural since no big ideas, no Faust keep him occupied. The doctor addressed him; unfortunately I was unable to hear him speak, but his smile and, so it seemed, his speech was completely as in the past. Herr Sch. then looked at the flowers and walked deeper into the garden, toward the beautiful vista. I saw him disappear as the evening sun surrounded him beautifully.*[*]

Does he ask about me? Clara asks in return, though she's already guessed the answer.

He loves her enough to release her.

He will not so much as utter her name.

---

* Litzmann, *Briefe*, 12.

What joyful news . . . that Brahms, to whom you will give my kind and admiring greetings, has come to live in Düsseldorf; what friendship!

—Robert (in Endenich), in a letter to Clara, 1854[*]

For the beautiful word in your last letter, for the love-felt "Du," I have yet to thank you most cordially; now your very kindhearted wife has also given me joy by using the beautiful, trustful word . . . I will always strive more to deserve it.

—Brahms, in a letter to Robert (in Endenich), 1854[†]

I have you, beloved Johannes, to thank for the kind generosity you have shown my Clara. She is always writing me about it.

—Robert (in Endenich), in a letter to Brahms, 1854[‡]

---

* Swafford, *Johannes Brahms*, 119.
† Litzmann, *Briefe*, 61.
‡ Ibid., 37.

16.

Hart called at the end of May, on the Friday night before Memorial Day weekend. As soon as I heard his voice, I understood I'd been waiting for this call, expecting it like news: good or bad, it was impossible to say. Over a month had passed since our dinner at the Wine Cellar. My academic semester had ended. For the next six weeks, I could write from eight until two, the hours Heidi went to Montessori school. It seemed like a lot of time. It wasn't. Especially since I was still stumbling, fumbling, searching for whatever was missing from the story I wanted to tell. On the twelfth of July, Heidi's summer session would end. On the thirteenth, Cal would pick her up for a two-week vacation—the two weeks I'd be in Leipzig. Maybe he'd take her to a reenactment somewhere. Maybe they'd just stay home. Or maybe he'd take her only one week, not two, in which case, as he well knew, I didn't know what I'd do.

The manuscript was due on August eleventh, two weeks after I came back from Germany.

"May I call you back?" I said, glancing at Heidi, who was busy unrolling a strip of butcher paper over the craft table. This weekend belonged to Cal as well. He was late, so I'd been distracting us both with stamps, stickers, finger paint. Last year, he'd taken a job at a private school in Lakeland, two hours to the north. Perhaps there'd been an accident and he was caught in a jam on I-95. Perhaps he'd gotten a late start. Perhaps he was about to pull into the driveway. "We're waiting for my daughter's dad to pick her up."

"Sure, sure," Hart said. "I suppose after that you have plans?"

His accent was more pronounced than I remembered. It sounded like disappointment.

"I'm meeting a friend for dinner," I said. "I haven't really thought beyond that."

"He gets his weekends with her, I suppose," Hart said, and it took me a moment to realize he was talking about Cal.

"Alternating. Same with holidays."

"My daughter lives in Paris. I saw her in London last week."

"Your ten-year-old," I said, pleased to display one of America's scant facts. "What was she doing in London?"

"Friederike is sixteen," Hart said. "Did those stupid people tell you ten? How hard can it be to get these things right?"

"I'm ready to paint," Heidi announced.

"Oh well," he said, relenting, "ten years, twenty, what does it matter? Perhaps she comes to the U.S. to study. In New York City. Have you been to New York City?"

"Mommy?"

What I said to Heidi: "Mommy is on the phone."

What I said to Hart: "I used to live there. I wish I still did."

"You could move back."

"No, it's too expensive, and besides, Cal's here, in Lakeland, and I have this tenured job—" Heidi had managed to open the green; she plunged in both thumbs. I trapped the phone between my jaw and collarbone, reached for a paper towel. "Look, I really can't talk right now. Sweetie, let me help with that, okay?"

"Your ex-husband is always late, I suppose. To bug you. It's the way these things go. Do you want to come flying with me tomorrow?"

"I—can I call you back? I need to think about it."

"You know the Starbucks off PGA? I can pick you up at eight."

"In the morning?"

"I will bring good muffins. From the Whole Foods. And fruit."

"I don't know."

"You are not liking fruit?"

"I'm not liking flying. I mean, I just don't know, I've never—"

"A lit-tle before eight, perhaps?" he said.

"Eight," I said firmly. "Let's have coffee together. Then we'll see."

"Okay, okay. Ciao."

I hung up, attempted to concentrate on the task at hand: red stars, a fat yellow moon.

"What color do you want the sky?" I asked Heidi, who was hard at work making fat blades of grass.

"Not blue," she said, "and not green."

I dumped out a little puddle of green, added a dollop of shining blue. She wiped her hands on a paper towel, hesitated.

"Can you do it?" she said.

I made small circles with my index finger, working the color over the page. The surface of the craft table was the same yellow as the moon, so when I finished, the moon looked like an absence, an overlooked space. I wanted to color it red, to match the stars, but Heidi disliked this idea.

"I want it to be lonely," she said.

Lonely: the word she still uses to mean *different.*

It was after eight by the time Cal arrived. While I packed up a juice box and crackers, he opened the refrigerator, the way he would have done when we were married. He took the jar of peanut butter, scooped, swallowed thickly. He considered the open bottle of Chardonnay in the door. There were leftover meatballs on a plate; he popped one into his mouth. Everything about the way he stood was daring me to tell him not to do this.

"I don't want to go with Daddy," Heidi said, though she did. And didn't. I knew exactly how she felt.

"I'll be right here when you get back," I said, pulling her into my arms for a kiss.

Her bags were packed. She was dressed. She wore shoes. Still, Cal was standing before the open fridge as if caught in the light of a shrine. Another meatball. A paper-thin slice of lox. A handful of blueberries. How could I ever have filled such emptiness?

"Calvin?" I said. "We're ready, okay? Take her, okay? Please."

17.

AT SCHOOL, HEIDI SPENT an entire day working on a drawing of her family. Her teacher labeled each stick figure according to Heidi's directions: Grandma Joan, Granny Hobbins, cousin Kayla, cousin Ray. Fourteen stick figures in all, each of them smiling, even the one named Heidi. All except one, the largest of the figures, positioned at the center of the page.

This figure is frowning. Her frown is colored brown.

"How do you spell 'Mom'?" Heidi had asked.

Then she'd labeled me in her own unsteady hand.

"Why is your mommy frowning?" the teacher had wanted to know.

(She'd tell me about this later, in a parent-teacher conference.)

"Because she is lonely," Heidi said. "My mommy likes to be lonely."

I am learning to understand Johannes's rare and beautiful character better every day. There is something so fresh and so soothing about him; he is often so childlike and then again so full of the finest feelings . . . And as a musician he is still more wonderful. He gives me as much pleasure as he possibly can . . . and he does this with a perseverance that is really touching . . .

—Clara, in a letter to Joachim, 1854*

He told me much about himself, which half fills me with admiration for him, half troubles me . . . Will not those who understand him be few in number?

—Clara, in her journal, 1854†

---

* Nora Bickley, ed. and trans., *Letters from and to Joseph Joachim* (London: Macmillan, 1914), 71–72.

† Berthold Litzmann, *Clara Schumann: An Artist's Life*, vol. 1 (New York: Vienna House, 1972), 81.

18.

Ellen is older than I am, dark-eyed and reckless, with a grin that suggests all kinds of good mischief. After her last marriage ended, she moved to Florida to be closer to her sister; now she manages investments for high-end clients at a private bank. As I hopped up onto the high stool beside her at the bar, it occurred to me, not for the first time, that we might be poster children for the maxim *opposites attract*. She sparkled in a white silk blouse, pink lipstick; I wore my usual jeans. At the bottom of my purse was the Chap Stick with sunscreen—sandy from my last trip to the beach—I'd remembered to swipe across my lips before getting out of the car.

"Got any plans for the weekend?" she said, leaning over for a kiss.

She'd already ordered wine for us both, gotten our names on the waiting list. The buzzer rested between us, red eye blinking, an old man's wink.

"Actually," I said, "remember the German entrepreneur?"

Ellen raised one shapely brow. "The one who says men and women can't be friends?"

"He just called me."

"So where's he been all this time?"

I shrugged. "He asked me to go flying with him tomorrow."

To my surprise, Ellen laughed. "Of course he did. Every other guy I've been out with lately either has his pilot's license or wants his pilot's license."

"I thought he was being original."

"*Men and women can't be friends.* That's original, all right."

The waiter brought our wine; I took a sip, leaned back. The bar was under a pavilion overlooking the intracoastal, which looked pretty the way an artificial Christmas tree can look pretty sometimes, despite the garish tinsel, the overreflected light. It was a pleasant place to sit and talk, to admire passing boats. On the opposite side of the seawall, ibis roosted in the trees, readying themselves for the night.

"Of course," Ellen said. "It's also the truth."

"Oh, God. Not you, too."

"Wait till I tell you my latest online success story. Are you writing these things down? This stuff is too good to waste."

"You mean Ketchup Man?" Ellen had met Ketchup Man on eHarmony.com. They'd e-mailed for weeks before agreeing to meet at a Subway, where—his suggestion—they'd ordered sandwiches to go. "Do you like ketchup?" he asked, leading her across the parking lot to Publix, where he purchased a family-size bottle. "They never give you enough ketchup on these things." Ellen confessed she did not like ketchup. "More for me!" Ketchup Man said. By the time they'd finished eating, sitting in his car, the bottle was almost empty.

In between squirts, he'd licked the cap.

"Not Ketchup Man," she said. "Dancing Man. Who, by the way, was a certified flight instructor."

"When he wasn't dancing."

"Or cheating on his wife."

I looked at her and saw she wasn't smiling. "What happened?"

"I met him a couple of months ago. Remember I told you I signed up for this dance class? Actually, we signed up together."

"Did you tell me about him and I just forgot?"

She shook her head. "I didn't want to jinx it. I mean, we had so much *fun.* And he seemed—well, we'd go out for dinner, dancing, whatever, and then we'd say good night and go home. But this past weekend was our last dance class, and there was a contest that, of course, we won. So afterward, we've got this ridiculously huge trophy, and I say, *Where should we keep it?* And he says—wink, wink—*Your place or mine?* Well, it's the first time that *this* has come up, and it's perfect, it's just right, so I tell him, *Your place,* because mine's a mess, and we jump in our cars and off we go. As soon as we're in the door, he's pulling me toward the bedroom, fine, but I really have to pee, so I twist away into the bathroom and I'm sitting on the toilet when I see—hello?—a box of tampons by the wastepaper basket. He's outside tapping on the door, and I'm like, *Just a minute,* but by now I'm opening drawers, and there's all this makeup, hair ties, nail polish, and that's when I find the birth control pills. Prescribed to a woman named Linda. Coincidentally, they have the same last name."

I didn't know what to say.

"So then I look in the laundry hamper and, I kid you not, there's this lacy scrap of a nightgown in there. I think about putting it on, but in the end I just throw it over my shoulders like a goddamn boa and open the door. He sees it—I mean, you can't miss it—but still he's ready to go. So I say, *Are you married?* And you should have seen his face. He turns bone white, and get this, he says, he says—"

She was laughing now, despite herself—

"He says, *How did you know?*"

"Jesus," I began, but Ellen said, "No," and picked up her glass of wine. "Don't say anything, okay? Just write it down."

She looked past me into the darkness. A sport fisher passed too fast along the intracoastal, disregarding the no-wake zone, and all the little runabouts tied along the pier bumped roughly against each other. Suddenly, I was wishing I'd answered L—'s e-mail, wondering if it was too late to do so. I wanted to ask him if it ever got easier: this dating, this dodging, this longing for love. The second marriage. The second time. I simply could not imagine it.

When the buzzer went off, it startled us both. A waiter appeared to help carry our wine.

"What do *you* think?" Ellen asked him, as if he'd been

part of the conversation all along. “Can men and women ever be friends?”

He was somewhere in his early thirties: tall, dark, and handsome. His teeth shone like something you could spend.

“When a woman asks that question,” he said, eyeing Ellen’s full chest pleasantly, “there is only one answer a man can give.”

Date: Friday, May 26 11:56 PM

To: LMJPROF@que.edu

Hey there—

Sorry I didn't write back sooner, but you asked how I was doing, and I've been trying to figure out how to answer. Maybe the old joke sums it up best:

Q: Why do people pay so much for a divorce?

A: Because it's worth it.

Congratulations on your new marriage—what is that like? I wish you both the best.

Take care,

Jeanette

Date: Saturday, May 27 12:02 AM

To: Jeanie88@comster.com

Oh, no, Jeanie, are you cynical now? Be careful with yourself. Heartlessness comes next. Write me a real letter if you can, when you can, and let me know (really) how you're doing.

You have always been special to me.

L—

Perhaps a primary reason that women are often so shallow and senseless is exactly their superior talent for the external. One sees with what comical verisimilitude little girls play brides, wives and mothers. . . . Many grown women experience romance and carry it off with a convincing show of sincerity, but in fact it is nothing more than a reiteration of their children's games . . . until in their imaginations it is as if they have really felt passion . . .

—from Brahms's notebook; copied from Friedrich Sallet's *Contrasts and Paradoxes*[*]

* Swafford, *Johannes Brahms*, 121.

19.

Hart was already in front of the Starbucks, idling in his Mercedes, by the time I pulled up beside him in my Volvo with its Cheerios décor: a high-backed car seat, Winnie-the-Pooh sunshade, Styrofoam noodles, and a half-inflated dolphin pool toy filling the back hatch. A couple of sagging helium balloons trailed me out the door, but I saw them in time, beat them back with my sneakers as Hart opened his own door to greet me. He wore a cloth hat, long pants covered with pockets, and a long-sleeved flannel shirt, despite the sun, which was hotter than you'd think possible at a quarter after eight in the morning. There was a funny moment when we looked at each other, and I saw that we were both disappointed. Regardless, we shook hands like proper Germans, and he said, "Did you bring a hat?"

"Do I need one?"

"And sunscreen."

"I always wear sunscreen," I said, but he was studying my tank top, my bare arms and shoulders.

"It is better to wear a shirt. What about your eyes?"

"What do you mean?"

He stepped forward, peered impersonally, clinically, into my eyes. His own eyes were close-set, somewhere between gray and green. A few wiry hairs escaped the arch of his brows. I wanted to touch the faint scar on his forehead. I wanted to push him away. Again, all of this seemed familiar, as if we'd stood like this before, scrutinizing each other too closely, looking for something we were not going to find.

Ellen was right: too much didn't add up. Why had he waited this long to call? And then why these last-minute plans?

"You should be wearing sunglasses," he said.

I stepped back. "So should you."

"I have seen many cases of melanoma."

"In your research, I suppose."

"I no longer involve myself in research. Your eyes, by the way, look good. Based on what I can see. Which isn't much. You should have them checked."

And why were we still in the parking lot? Something was off between us, out of step. We'd have a fluffy coffee, and then I'd send him on his way. Get my laptop from the car. Spend the morning doing what I *should* be doing: writing to the sound of the baristas. Maybe today would be the day an overlooked detail would open a door into a room yet unimagined. Into my passion for these people, this story I had loved for almost thirty years.

"So what *do* you . . . involve yourself in these days?" I asked. "Your business, I suppose. Is Viso-Tech a large company?"

"Sure, sure, I am the rich businessman. We can talk while I am driving." He nodded toward the car. "It is a long way to get there."

"I thought we were going to have coffee first."

"I bought lattes." He opened his door and there they were between the seats: whitecapped soldiers in wrap-around jackets. The sight of them unnerved me. In the back, there was a fat padded cooler, a half-zipped satchel stuffed with papers and books, a neatly folded blanket. Everything set to go. It occured to me that this time the voice in my head crying warnings might be right. Why hadn't I left his phone number with Ellen? Why hadn't I thought to make certain Viso-Tech really existed?

"I never even asked which airport," I said.

Hart got into his car. "*Glider* port. It is west of Orlando."

"But that's over three hours away!"

He looked up at me inquiringly. "You must be back by a certain time?"

The Mercedes ran so quietly I didn't even realize he'd turned the key until I felt the first cool puff of air-conditioning. "Look," I said, taking a few steps back. "I don't even know you. It's too far. It's too much."

Hart did not say anything.

"Even if I did come along for the ride, I don't think I could fly. I'd be too afraid. I mean, I *am* afraid. Of everything, these days." I made myself look into his face as I said this. "Not just you. Not just this."

He said, without missing a beat, "I do not find, since I am living here, so many people I can talk to. I was thinking that perhaps we are two people who can have a conversation."

"So call me sometime. We can talk on the phone. Get to know each other better."

"Sure, sure."

"It's the way these things go. You told me that, remember?"

He traced the downward curve of the steering wheel.

"Only then," I said, "you didn't call. You disappeared. Is that also the way these things go?"

He said, "There is a certain chemistry that must exist between a man and a woman. I am thinking this chemistry does not exist between us."

Ibis threaded their way along the narrow strip of grass that divided the parking lot from the sidewalk. I waited to feel something: embarrassment, maybe. Disappointment. But nothing in particular came to the surface, other than the feeling that we'd already had this conversation. That we were just pretending there was a question on the table, a decision to be made, when in fact it had already been settled.

At last I said, "The chemistry is more like . . . *murky* . . . don't you think?"

"No-no-no." He looked at me, unflinching. "For me, it is not there. When I think of the sort of person with whom I wish to be involved, I am sorry, she is nothing like you."

"Fine," I said. Oddly enough, it was something of a relief, knowing he wasn't looking at me as a woman. Or at least not a woman he planned to date. "I mean, it really is fine." I wiped the sweat from my face and neck. "I'm new to all this anyway."

"I know. I am not new to this."

"I know."

"And yet, there is something about our first conversation. It is difficult to explain. After we met, I arranged to see my daughter in London. I stayed there two weeks. She is a violinist, did I mention this? And she knows all about your Clara. There's a house in Leipzig where she and Robert—"

"On Ingelstrasse?" I stepped forward again. "It's a music school now."*

"I must have passed it countless times."

"I'd love to get inside it, but I'm not sure it's open to the public anymore."

"It is open for concerts. Friederike has plans to perform there." A little smile played around his mouth. "It would seem we have another coincidence."

She was, in fact, scheduled to perform on the evening of the day I arrived.

"Friederike will get us tickets. I mean, if you would like that. I am happy to show you around, to be helpful to you. If I may."

"As a friend," I said, understanding him.

"I think, yes, as a friend," he said, nodding. "I like to talk to you. I have told you that already."

"Men and women can't be friends," I said. "You already told me that, too."

---

* www.schumann-verein.de.

"I made you very angry when I said that."

"You did."

"I'm afraid I am still thinking this is true."

I started to laugh, I couldn't help it. "Are you always this complicated?"

"If we were to leave right now," Hart said, "I'd have you back by eight. Plenty of time left to chop up your body and bury it deep in the ground."

I grabbed my computer case from the floor well beneath Heidi's car seat, added a short-story anthology I'd been meaning to read. The inside of the Mercedes was leather lined, cool, the color of heavy cream. He was right, I decided, about the chemistry. How could there be chemistry when it was suddenly this comfortable, this easy? No resistance to chafe the match. No rough edges to spark. Getting into this car, sliding into this life, was like continuing a conversation we'd already begun. The smell of the interior was familiar as bread. I recognized the coins in the cup dish, the zippered case for CDs. The beaded bracelet hanging from the rearview. The cubby for the mirrored sunglasses he removed carefully from their sleeve.

"I do have sunglasses, you see," he said, concealing his eyes behind the reflection of my own.

"Pretty bracelet."

"It is something belonging to Friederike. Take this. No, this one."

"Is it sweetened?"

"A lit-tle."

"I don't like sweet coffee."

"It is the only way to drink this fucking American coffee."

"My God, are you always like this? What's wrong with American coffee?"

"It tastes like dishwater. It tastes like such strip malls you see everywhere. It tastes like these morally reprehensible high-rise developments."

"You just make it worse with sweetening."

"Fine, fine. Next time you shall have no sugar."

A chemical glitch, I reminded myself.

Together we buckled up for the ride.

Part IV

Blue Day

Schumann's Beethoven, 2006

I am always pleased by Beethoven's statue and the lovely view towards the Siebengebirge.

—Robert, in a letter to Brahms, 1854*

* Litzmann, *Briefe*, 37.

Oh, if I could only see you and talk to you, but the road is too far. I would like to know so much about you, what your life is really like, where you are staying, and whether you still play as marvelously as you did once. . . . Oh, how I would love to hear your beautiful playing.

—Robert to Clara, from the asylum, 1854*

I am haunted by music as never before; at night I cannot find sleep, and by day I am so absorbed by music that I lose track of all else . . .

—Clara, in her diary, 1854†

Eugenie seems to have caught cold, she has no appetite, her face is flushed . . . The boys are fine, even Felix. We do not progress much with the alphabet, despite large portions of sugar loaf . . .

—Brahms (in Düsseldorf) to Clara (on tour), 1854‡

---

* Bertita Harding, *Concerto: The Glowing Story of Clara Schumann* (Indianapolis: Bobbs-Merrill, 1961), 134.

† Ibid., 133.

‡ Litzmann, *Letters*, 33.

20.

THE FLORIDA COASTLINE GAVE way to its drained interior, flat, fenced fields fringed with horned cattle, gladiola farms, tomato farms, crossroad towns with their concrete-block churches and faded American flags. Hart talked about agriculture, about industry, about the energy crisis. He talked about the Florida election scandal. He lamented the religious right, religious belief in general, its distortions, great and small. He talked about racism, homophobia, the hopeless situation of the working poor. He reminded me of a man who hadn't spoken for years, the result of an illness or curse, who must say everything as rapidly as possible before he'd be silenced again.

At one point, his cell phone started to ring; he answered in English, switched to French.

Écoute, je ne peux vraiment pas te parler maintenant.

Non. Non.

Il faut que je te laisse. Je te rappelle plus tard.

"How many languages do you speak?" I asked after he'd hung up.

"Oh, the usual," he said. "German and Russian. English. French. Some Spanish. I am always working to improve my Japanese."

"Ah," I said.

We passed a church billboard that read, YOU CAN'T HIDE FROM GOD BY MISSING CHURCH.

"You can't hide from Satan by going to church," Hart said.

"You believe in Satan?"

He avoided the question. "*You* believe in God, I suppose."

"Yes and no."

"And now I am waiting for the long explanation." He selected a CD.

"I don't believe in any external entity," I said, kicking off my sneakers and settling my feet on the dash. "What I do believe is that so many *others* believe that the sheer block force of it has an impact on everyone. Belief in God is the reason, for example, why kids here in Florida public schools are taught creationism."

"This is a joke."

"I'm afraid it's not."

"America: land of the free, home of the religious right. Must you put your feet there?"

"I must. You'll have to deal with it."

"It is causing me physical pain."

"They're clean. And in case you haven't noticed, *I'm* an American, and *I'm* not preaching creationism."

"No, you are just a nice Catholic girl."

"Actually, I'm a practicing Catholic. I practice and practice but never get better."

"That is funny." The CD had kicked in, a soft piano introduction. "So you believe not in God," Hart said, "but in the *significance* of God."

"Hey, is this Clara Schumann?"

"You are awfully fond of that word: *significance*." He turned up the volume. "Perhaps you rely on it too much."

"The G minor Trio!"

He was pleased. "Friederike recommends this CD. The trios of Clara Schumann and Fanny Mendelssohn."

"Felix Mendelssohn's sister, sure! A wonderful composer. By the way, Clara and Felix Mendelssohn were also friends. In fact, Clara named her last child Felix. At Robert's suggestion, no less." I glanced at him. "Though I suppose you will tell me that this suggests Clara Schumann and Felix Mendelssohn had an affair."

"A significant possibility."

"Seriously, did you know she had another lifelong male friend? The violinist Joseph Joachim. Who was also, incidentally, a lifelong friend of Brahms. Robert referred to them, once, as the two young demons."

"*Three* men having inappropriate relations with his wife. No wonder the poor sucker went mad."

We were laughing.

"It would be a fresh angle, that's for sure," I said.

"*Ja,* but try to publish a story like that in America," Hart said, and suddenly he was serious. "Religion in this country is worse than communism. Say the wrong thing, think the wrong thing and, look—there's the Stasi knocking at your door. I suppose you will say I exaggerate, but when I think of what's happening here, it terrifies me."

We were more than five years into the Bush administration.

I could not disagree.

Heat waves shimmered over cracked sidewalks, rose from tar-pitch roofs. We passed smashed armadillos, smashed tortoise shells. A black leather ribbon of desiccated snake. Just beyond Yeehaw Junction: seven white crosses all in a row, each with its faded memorial wreath. High overhead, the sky stood watch: cloudless, vast, uncompromising.

"Blue day," Hart said, and then he sighed. "Hard to find lift."

"Lift?" I said, and he pointed to a lazy spiral of vultures: five black flecks above an irrigated field, drainage canals dry as rust. "What they've found is an injured animal," I said, but he shook his head.

"They are riding a thermal. A column of air. Warm air rises, yes? There's your lift. Usually you look for cumulus clouds, the kind—you have seen them?—with those dark, flat bottoms. On a blue day like today, you have to look for birds. Or another glider. Or else you go by the way things feel. Intuition, if you will."

"How do you find your way back?"

"Visually, for the most part. Though there's also GPS. And I always carry a chart, just in case. Here, I'll show you." He reached behind him, steering with his knees, and pulled his flight bag into my lap. "Look in the side compartment," he said, but I was distracted by a copy of *Clara Schumann: The Artist and the Woman* by Nancy Reich, flagged with brightly colored Post-its.

"What's this?" I said. First the CD, now the book. It was a little unsettling. All this effort for a woman who inspired no chemistry?

"Yes, I have been reading about your Clara," Hart said. "She is compelling, this is true. As a woman, as a person. Friederike is sending me *Erinnerungen*—"

"Eugenie Schumann's memoir? I've read it in English, but I can't—"

"Perhaps we could look together at both editions. I could also translate letters or entries from diaries—"

"I was going to ask you about that, but it seemed like a lot—"

"It is something that interests Friederike. Something she and I can talk about. It is challenging, when one lives away from one's child, to find a common ground. Especially as time passes. But you are not telling me the whole story, I think, when you speak of Clara's friendships

"Don't you two get along?"

"What are we talking about?"

with Mendelssohn and Joachim. Or perhaps you do not understand the *significance* of a man the age of Brahms, living in his time, calling a married woman, an older woman, not to mention a great pianist and public figure, *Du.** To me, it indicates there was more than—

"But I *do* understand—"

"I understand that *Du* was usually used—"

simple friendship between them."

"—*usually* used between spouses or sweethearts, family members."

"*Exclusively* used—"

"But Clara said she felt like a mother to Brahms. She loved him, yes, but like a son."

"*Would to God that I were allowed this day instead of writing this letter to you to repeat to you with my own lips that I am dying of love*

* The informal you pronoun *Du* was used almost exclusively among family members, husbands and wives, and those engaged to be married. Lifelong friends, even of the same sex, age, and social status, used the formal *Sie* when addressing one another.

for you. Tears prevent me from saying more.[*] *Ja*, that is one devoted son."

I stared at him. "You have memorized Brahms's letters?"

"It is nothing I set out to do. It happens on its own, I cannot help it."

"Have you always had this kind of recall?" I asked, but the question didn't interest him, and he waved it away like smoke.

"Should I say such things about my own mother, I believe they would arrest me, these Stasi religious American pigs."

"Um, yeah, but you have to take into account," I said, recovering my footing, "the fact that Clara would have been used to men of all ages fancying themselves in love with her. I mean, she was a phenomenon. *Das Wunderkind. Die erste Pianistin.* The men were in love both with the music she made and with the way they could speak to her *about* her music. Almost as if she were not a woman."

"Different," Hart said, lifting a finger, "than speaking to her as if she were a man."

"True."

"And none of these other admirers, even those of her own station, dared to address her as *Du.* Certainly she never used it. Except with Johannes."

---

* Litzmann, *Letters*, 20.

"But this, too, can be explained," I said. "There's a quote from her diary where she says that an artist is not to be judged by age, but by intellect—"

"And when I am with Brahms"—Hart provided the quote—*"I never think of his youth, I only feel myself wonderfully stirred by his power and often instructed."*[*]

"I wish I had your memory," I said.

"Don't ever wish for that."

"But I forget everything."

His cell phone started up again.

"And I," he said, silencing it, "cannot forget anything. Even when I try."

Another church billboard: WHEN GOD SAYS NO, IT'S BECAUSE HE IS GOING TO SAY YES TO SOMETHING BETTER.

What Hart said: "Such as Satan."

"You're not married, are you?" I asked.

He gave me an amused look. "At this particular moment in time? I am not."

"Do you think you'll get married again?"

He considered this. "Sure, sure. Probably. Won't you?"

"I would have to have a good reason."

It was the first time I'd heard him really laugh. "You did not have good reasons the first time? My, my."

I found myself laughing, too. "We thought it would be nice to get married," I said. "So we did. That's how young we were."

"It was the same with my second wife," he said, and

---

* Swafford, *Johannes Brahms,* 148.

I was about to ask exactly *how* many wives there'd been when his cell phone bleated. A text this time.

"Is there some kind of problem?" I asked.

"It is Friederike's mother. That woman will drive me crazy."

"What did you do?"

"What did *I* do? Sure, you are a woman, so you assume I've done something wrong."

"Did you?"

"I flew to London, where I saw my daughter, under the supervision of her teacher. I have spoken to my daughter, since, by phone. These are the recent crimes of which I stand accused. Friederike wishes to study in the U.S., as I have told you. At the Juilliard school. You have heard of the Juilliard school? She has been accepted there."

"Wow."

"It is not me putting such ideas in her head. Her teacher arranged the audition. Of course I do not disapprove."

"Her mother does?"

"Lauren herself will never leave Paris. Have I mentioned she is recently engaged? To a man, I might add, who has as little understanding of Friederike's gifts as Lauren herself. Both would be content for Friederike to remain at the conservatory for the next four years. Marry at twenty or twenty-two. Settle down, have a baby or two. Give up these foolish dreams of a concert career."

"All right, I'm sorry."

"No, no." His mood had already settled into weary resignation; he seemed, all at once, like a much older man.

"Lauren has her point. You know how it goes. The child is happy with how she lives. Then the father gets involved. After that, she and her mother are at odds."

I thought of Cal's rule, one I'd come to admit made sense. When Heidi was with him, she might ask to call me, but I was not allowed to call her. The intrusion only upset her, disrupted her, interrupted the life he was trying to make for her there. Still, I didn't know the answer to this: for Heidi, for Friederike, for anybody's child.

Hart and I rode in silence for a while, the absence of our daughters connecting us: as solid, as real, as any physical presence. Except that Heidi's absence was only for the weekend, a total of four nights out of each month. Except that, for me, that absence offered a guilty relief, a gulp of fresh air I could only imagine as soon as I was back in our daily life together: getting her off to school, preparing her snacks and suppers, cajoling her through piano lessons, tucking her into bed. What I wouldn't have given for a real stretch of time—not just a few weeks, but several months—to write the way I used to write. To immerse myself fully, completely. No need to come up for air every few hours, for hugs and kisses and story time, for fevers and board games, laundry and dishes, endlessly sticky countertops, endlessly sticky hands.

"Don't ever forget you are the lucky one," Hart said, as if he'd been reading my thoughts. "Remember, you are the one with the child. He is the one without her."

When I am able to practice regularly, then I really feel totally in my element; it is as though an entirely different mood comes over me, lighter, freer, and everything seems happier and more gratifying. Music is, after all, a goodly portion of my life, and when it is missing, it seems to me all my physical and spiritual elasticity is gone.

—Clara, in her diary, 1853[*]

We knew that in our mother woman and artist were indissolubly one, so that we could not say this belongs to one part of her and that to another. We would sometimes wonder whether our mother would miss us or music most if one of the two were taken away from her, and we could never decide.

—Eugenie Schumann[†]

---

* Reich, *Clara Schumann*, 287.

† Eugenie Schumann, *The Schumanns and Johannes Brahms* (Lawrence, Mass.: Music Book Society, 1991), 152.

21.

She's arranged for the coach to arrive before dawn, while the children are still sleeping. Better this way, Clara tells herself. Awake, they'll only cry and cling. Beg her to stay.

Like Johannes.

But what good can there be in sitting home when she might be performing, earning money for them all? Here she's no use to anyone, dissolving into tears at a word from a friend, at a letter from the doctor, at the sight of the Rhine. Better for her to get back on the road. Better for the children to be left in the care of stronger minds, less tremulous hands. And now that they're settled in the new Düsseldorf apartment—Johannes established in his own rooms below, keeping a daily eye on them—she can set off without fear. The housekeeper will take charge of the four youngest children until Ludwig and Ferdinand can be sent to school. Felix has been weaned, the wet nurse dismissed. The cook has decided to stay, thank goodness, and the new maid, though young, shows promise. Julie, always sickly, will be sent to her grandmother in Berlin. Marie will return to her Leipzig boarding school, along with Elise—though Elise is turning out badly: obstinate, unpleasant, indifferent. Hard not to fear for her future, for what kind of life can be in store for such a girl? Not unpretty but certainly no beauty.

No particular talent.

Then again, what kind of life can be in store for any of them, now that Robert's madness has been announced in newspapers as far-flung as America? Perhaps, the younger ones will show signs of his genius, but the oldest have disappointed her, she has to admit that this is the case, despite Marie's sharp, diligent ear; her discipline; her steadiness of character. How fortunate that Johannes doesn't mind the children's many shortcomings, listening to their chatter without the least sign of impatience or exhaustion. He's practically a child himself, entertaining them all with gymnastics and riddles, terrorizing them with the same monster games Robert once played with herself and her brothers.

Chasing them around the *Piano-Fabrik*.

Popping out—*buh!*—from behind closed doors.

How often she thinks now of that long-ago life with her father in which nothing was required of her beyond what she most wanted to do. Fresh sheets of music set before her like maps. Long walks for the sake of her constitution. New white dresses at the start of each concert tour. The surprise of new towns, unfamiliar performance halls, each piano like a human face, never to be forgotten.

Now, as she waits in the early morning darkness, she feels herself growing younger. Stronger. At the sound of approaching horses, she opens the door, hurries down the steps, almost expecting to see her father—but, no. Her maid, waiting with the lantern, is waving the coachman inside to fetch the trunk. And here is Johannes, sleep-rumpled, bareheaded, lifting the heavy satchel of music. He presses something round and hard into her hands—an apple wrapped in his own handkerchief—and though she doesn't want to accept it, doesn't want to be burdened with anything more to carry, she finds a place for it in the already bulging pockets of her travel cloak, actually Robert's greatcoat, oversize but warm.

Johannes, her friend. Her dear, true friend. Perhaps he is falling in love with her, but what can be done about that? She is, after all, still married. And if he chooses to show his love for her—for Robert, too—by staying with the children, she is not in a position to refuse such a gift.

"Kann ich denn nicht mit dir kommen?" he says, but she remembers too well how it used to be with Robert, who also begged to accompany her, only to sulk at the attention she received, nursing slights (real and imagined) until he worked himself into fever. No doubt Johannes will behave the same way. No doubt, at some point, she'll be forced to choose between him and the work she most loves. Already he's trying to convince her that performing is not important. He believes she should stay at home, live quietly. Compose.

If only everything could always be exactly as it is between them!

If only they could remain as they are—friends, best

friends—forever.

"*Mein Johannes,*" she says as he embraces her through the padded flesh of Robert's greatcoat. He is telling her something about the children, does she have any message for them? No, there is not any message. The children have everything they'll need. It is best that she leaves this way, in darkness. It is best that, by the time they wake to their bread and hot milk, she'll be gone.

"*Und hast du nicht wenigstens ein liebes Wort für mich?*" he asks, but she has no words for him, either. Her heart is with her husband. Her heart is like a stone.

It is only after the carriage is far beyond Leipzig, after the maid has fallen asleep, as Clara herself drowses—lulled by the rattle and bump of the road—that she feels for him, and for the children, too, that tenderness of heart her father once feared, fought against fiercely, believing it would lead exactly to this: an unhappy marriage. Too many children. Small, dark rooms in which they've already awakened, knowing their mother is gone. If she opens her eyes, she'll see them now, running bare-legged after the horses. If she listens, she'll hear them sliding through the windows, slipping through chinks in the rattling doors. She'll smell the crisp scent of the apple as they pass it between them hungrily, silently, hoping she won't awaken, for if she does, won't she only send them away, the way she usually does?

Können wir denn nicht mit dir kommen?

Can't we come with you, too?

She is reading their thoughts as if they are speaking.

Why not? Isn't this the privilege of a mother's love? And it is love she feels, for the moment, for them all, now that it's safe to do so. Now that they are only the children of dreams. She can love them completely—Johannes, too—with whatever is left of her heart.

22.

I AWAKENED TO FIND we were on a dirt road. Hart said, "Do you always sleep like that?"

"Like what?" I was looking at a group of toy planes, life-size, scattered across a wide, mowed field. Each rested atilt, one wing on the ground, the other extending into the air.

"Like a dead person. It is spooky."

"Sorry."

He was still shaking his head. "You should warn a man about such things."

We parked against a low wooden fence. A few hundred yards away, across a yellowing strip of close-cropped lawn, stood an octagon-shaped house with a wraparound porch overlooking a grassy airstrip. As we approached, I saw two Asian boys, teenagers in shorts and soft-brimmed hats, standing around a soda machine, gulping cans of

Coke. A small, brightly colored plane, hardly any bigger than a sports car, peeled out of the sky and rattled to a stop directly in front of us all. The pilot—a middle-aged woman—hopped down, mopping at her face with a long, loose sleeve. As soon as she saw Hart, she greeted him by name. So did the teenagers.

Hart responded in Japanese.

Next to the house there was an in-ground swimming pool, landscaped with tall bursts of flowers, and as I followed Hart up onto the porch, butterflies rose in a single, fluttering cloud, distracting me so that I almost didn't see what everyone else had turned to watch: a second plane moving smoothly down the airstrip, towing one of the toy-like planes on an impossibly slender rope.

"Not a plane." Hart corrected my thoughts. "A glider."

Already, the glider was lifting into the air, floating behind the towplane. Then the towplane was airborne, too.

"That is Midori," the tallest of the boys told Hart.

"Have the rest of you soloed?"

"Midori after eight hours only! I solo after fourteen."

"No more today," said the second boy, who had his Coke pressed to the top of his head.

"Tighten your harness next time," Hart said, grinning at him, "and you won't bump your head on the canopy."

"Thank you. Very helpful," the boy said sarcastically, but Hart was already turning away, climbing the steps to the house. Which, it turned out, wasn't air-conditioned, just cooled with noisy fans. There was a couch, a long conference table with chairs, a circular staircase leading up to a closed door. A man in his fifties bent over a computer screen, clicking on colorful maps. Another man rooted through the full-size fridge. There was a coffeepot, a sink filled with cups. There was a signed picture of Neil Armstrong in front of this very house. The pilot came in behind us, seated herself at a desk half hidden beside a display case crowded with hats, T-shirts, pilots' logs, coffee mugs.

"How is the lift today, Miriam?" Hart asked.

"Booming, if you can find it. Hey, Chuck, see what the cat dragged in."

A deeply tanned man in a cowboy hat had appeared in the open back door.

"*Wie geht's?*" he said, and he shook Hart's hand. "I see you brought a friend."

"Who isn't flying," I said quickly.

Chuck grinned. "You want a lift out to the hangar?" he said, and Miriam said, "It's too hot to be walking around," but Hart waved both of them off, no-no, so we walked back outside alone, crossing the cropped grass in the beating-down sun toward an outdated mobile home. The towplane returned in a smooth, purposeful swoop, chattering past the house until it reached the far end of the airstrip. A man emerged from the cockpit, dropped to the ground. Hart watched, squinting against the light. "Miriam's brother," he said. "Both he and Miriam could fly before they could drive."

"What about Chuck?"

"He started gliding when he married Miriam. The three of them have been running this place for, oh, fifteen years or so."

We'd reached the mobile home. A single air conditioner perched whimsically in one of the windows, buzzing, dripping. "They all live here?" I asked.

"No, that's just the pilot's dorm," Hart explained. "Ten dollars a night. It is actually quite homey. Three bedrooms, a pull-out couch, satellite TV. There is even a piano."

"You've stayed here?"

"Last weekend, in fact. I got to practice my Japanese."

"These kids are *living* here?"

"It is cheaper to get licensed in the States. Kids always pick it up quick. That Midori, though, she is something." Hart whistled beneath his breath.

"Why do these kids need licenses?"

"You must have a license to fly."

"I mean, why do they need to fly?"

Hart shot me a puzzled look. "Why do you need to write books?"

On the other side of the mobile home stood a large metal hangar and, in front of it, a scattering of tethered gliders, each protected by a snug canvas cover. Beyond them, set on a concrete slab and sheltered by a wooden roof, were slim, rectangular trailers that, Hart explained, held disassembled gliders. His own, an ASW 27, was already assembled in the hangar; I was relieved to see it seated only one. Even a glance at the narrow cockpit made me feel short of breath.

"Want to help?" Hart asked.

"If I can do it from the ground."

"There's nothing to be afraid of. I have flown over two hundred hours."

"Life's too short," I said, pleased to deliver Ellen's smug line, "to get killed falling out of the sky."

Hart was not impressed. "People die all the time for many reasons. You would rather slip and fall in the bathtub?"

"I would rather live a long and healthy life."

"You do not have that choice."

"Yes and no."

He shook his head. "A significant answer. Pick up the wing, do you mind?"

He showed me how to hold it, and together we rolled the ship out of the hangar and pushed it across the sun-scorched grass. Fire-ant nests were scattered everywhere; I was glad I'd worn functional shoes. The end of the airstrip was a good quarter mile away, capped by a stand of pines, the ground rutted and scarred by wild pigs.

"I can take you up later in the Blanik," Hart said. "It seats two. They use it for training."

"Thank you, no."

"If you don't like it, I'll bring you right back down."

I sighed. "Which one is the Blanik?"

He pointed with his chin. The glider we'd seen earlier was descending now, skimming just over the surface of the airstrip to land in a hiss on the grass. A Japanese girl climbed out of the front, looking all of twelve years old. "Midori!" Hart called and she waved to us both, beaming, so beautiful I could not look away. Together we watched as she circled the Blanik, still hopped up from her flight. Already, she was visualizing her next takeoff. Already, she was resenting the lost time on the ground.

"I admire her passion," Hart said. "I used to be such a person."

"About flying?"

"About everything." Chuck, still in his cowboy hat, had sped from the house on a golf cart, and now he and Midori rolled the Blanik to the side, making room for us to pivot the ASW into position. Hart pressed his car keys into my hand, his own hand surprising me, cool despite the heat. "Here. In case you want something to eat. Bottled water. You are welcome to whatever you find."

"I'll probably grab my computer and write."

"Your book is going well?"

The question took me by surprise. Perhaps this is why I answered it honestly. "Not really. No."

"What is the problem?" His face was close to mine.

"The relationship was so contradictory. The facts just don't add up. Not that I'm trying to write a nonfiction book, but when a novel is working, all of it becomes true somehow, whether it is or not. A really great novel gets at the truth the way nonfiction can't."

He smelled, very lightly, of shampoo and sweat. Unabsorbed sunscreen clung to the stubble along his jaw. "Do not be offended," he said. "But why are you so opposed to the idea that their relationship was sexual?"

"It's not that I'm opposed," I replied, startled by his frankness. My heat-flushed face stared back at me, bug-eyed, from the lenses of his sunglasses. "But we have so few documented examples of straightforward friendship between great men and great women."

"You love the idea of such a friendship, I see," Hart said. "But perhaps your belief in such an idea says more about you than the people in your book." He touched my cheekbone; I flinched, then felt silly when he showed me the eyelash stuck to his finger. "Of course, that could also be part of the story. I would be interested to read such a book."

Before I could reply, Chuck pulled up in the golf cart, Miriam riding shotgun. "Ready to go?" she said.

I stood back to watch as she taxied the towplane into position. A long rope trailed behind it, which Chuck attached somewhere beneath the nose of the ASW. Hart was already in the cockpit, going through his preflight. At his signal, Chuck raised one of the wings, so both paralleled the ground. The rudder waggled from side to side, this time a signal to Miriam, and now both ships were in motion, gradually picking up speed. Chuck ran with the wing for five paces or so until the ASW was balanced, rising. The towplane, too, lifted into the air, the exacting distance between them an act of pure choreography. I watched until I couldn't see them anymore, shielding my eyes against the heat and the light.

Such an idea says more about you.

I didn't want to know what he meant. And yet I already knew.

Chuck and Midori picked me up in the golf cart as I

headed back toward the office. "Sweet takeoff," Midori said, smiling. "You are pilot, too?"

"Not me."

"You fly with Hart?"

"Later. If I don't chicken out."

"I'd be afraid to fly with him, too." Chuck waved his hat at a persistent buzzing fly. "These big guys, they're all the same. Risk takers. Buzz junkies."

"Hart's a big guy?" I said.

Chuck gave me a look of surprise. "Eye man. Surgeon. Figured out some kind of prosthetic eye. Damn thing works with the brain somehow, actually senses light."

"He developed idea in Japan," Midori said with evident pride.

"Company went public in the nineties," Chuck said. He gave a little whistle. "Beaucoup bucks."

"I guess I don't know much about him," I said.

"Tell you what," Chuck said, settling his hat back onto his head. "You got a nail sticking out of one eye? That's the guy you want to see out of the other."

Date: Saturday, May 27 12:16 AM

To: LMJPROF@que.edu

Dear L—,

You know the one about the old man whose grandson is getting married? Just before the wedding, he calls the boy in for a chat. "My child," he says, "I want you to know that all marriages go through phases. At first, you and you wife will make love all the time. But then, as the children come along, you will find that you are having sex less and less. And by the time they are grown and gone, you'll be just like your grandmother and me. All you'll ever have is oral sex. I just wanted you to know how things will go."

The boy looks at him, incredulous. "You and Grandma have oral sex?"

"Every single night," the old man says, "and it's a perfectly natural thing. She goes into her bedroom and calls, 'Fuck you!' And I go into my bedroom and call back to her, 'No, fuck you!'"

You are wondering when I got cynical? I am wondering when you got so fucking serious.

Jeanette

Date: Saturday, May 27 12:19 AM

To: Jeanie88@comster.com

Easy, Jeanie. I'm finding it hard to laugh at anything right now. Sally and I are separated. You get divorced once, okay, but twice?

L—

23.

I GOT MY COMPUTER out of Hart's car, grabbed muffins and fruit and two bottles of water, then set myself up at the end of the conference table, directly in front of the largest fan. Two hours later, I'd hacked my 230-page manuscript down to 50-odd pages. Gone: all the precious, meticulous details about German life in the 1850s. Gone: the clever subplots involving Clara's musical contemporaries. Gone: the carefully researched nuances of individual performances. Food preferences and wardrobe descriptions. The exacting layout of the Dresden house. All of it—procrastination, literary filler. Beautifully written ivy disguising a crumbling structure, empty rooms.

Facts that had nothing to do with the truth.

There are things about men and women that do not change.

What was left: Clara's backstory. Robert's, too. The first meeting of Brahms and the Schumanns. The moment when Clara finally understands her marriage cannot continue. *I am not worthy of your love.* And then—an indulgence, a pure waste of time—I wrote the same moment as it occurred between Calvin and me. As soon as I finished, I deleted it. There are things that don't belong on any page. I wrote, instead, a new scene in which a woman stands at the corner of Ponce and Dixie, unable to step forward, unable to turn back. There were the vultures I'd just seen, the clear sky overhead. There was my university, the corner where I cross the street to work, just blocks from the corner where a student was actually killed. Concrete, crosswalks, modern-day traffic. Nothing I was writing could possibly appear in a novel set in the mid-nineteenth century—

But wasn't this something I'd learned again and again? You went where the writing took you. You followed a serendipitous path. Perhaps what was missing from what I'd written was exactly this bridge between present and past. What could I take from the life of Clara Schumann as a working artist, living in the world today? As a mother? As somebody's former wife? As somebody standing on the edge of what must be a whole new life? Perhaps what was most remarkable, relevant, about the lasting friendship between Clara and Brahms was not that the two were never lovers, but that, indeed, they had been.

Once I would have said that Cal and I were friends.

Pilots came and went, some spreading charts beside me. The Japanese students appeared in pairs, studying for the written portion of their test. It felt right to be sitting among them, working in the company of others who shared—if not my passion, *a* passion. In the end, wasn't all of it the same? When Midori sat down with a notepad and pen, eager to practice her English, I stopped working to answer her questions. She asked me what I was writing, and when I told her, she giggled and covered her mouth with her hand.

"You write about me," she said.

"Maybe I will," I said. "You have such an obvious passion for flying."

Puzzle lines creased her smooth forehead. I tried to rephrase.

"Why do you love to fly so much?"

"Ah! First glider, then plane. My dream is earn commercial license, be pilot. Big plane."

"What I mean," I said, "is why do you love *flying* and not, say, ice skating or soccer or playing the piano?"

She shook her head, confused. "I play piano. Maybe I am very good at."

"But is the piano also your passion?"

Another puzzled look.

"Okay. If you had to choose between playing the piano and flying, which would you choose?"

"I practice every day. I made promise to parents," she said. "What is meaning passion?"

"Love," I said, "but stronger. Better."

"Please write down." She extended a little notepad.

I wrote the word *passion*.

I wrote, *What makes you passionate about flying? What if something happened so you couldn't fly anymore?*

I wrote, *What do you think makes some people want to do what you're doing and other people, like me, afraid to do it?*

Midori received the notepad from me with both hands. "I will look up in dictionary," she said. "I will answer all questions. For your book."

She bounded away. I picked an old scene.

Began to revise.

24.

It is not just his face that Clara finds beautiful, though she recognizes, there, what others see. Recognizes, too, his awareness of that beauty, his assumption that others will respond to it and treat him accordingly. Robert, one hand held to his mouth, cannot stop touching Brahms with the other, rubbing Brahms's smooth cheeks in greeting, clinging to him when he leaves. Of course the children cling to him too, the boys fleeing in mock terror from his kisses. And when he plays for them in the evening hours, she and Robert clasp hands as they did years ago, all they once felt for each other kindling, again, into shining flame. What genius. What passion. The mere thought of Brahms's boyish face, those soft hands and slender shoulders, strikes like flint against the rich, shuddering sound.

The children lined up on the couches to listen. Passersby standing outside in the cold.

It is said they remove their caps and scarves so as to better catch each note.

His outbursts of arrogance startle everyone, of course. "Brahms is ego incarnate, without himself being aware of it," Joseph Joachim will write to a friend. "The way in which he wards off all the morbid emotions and imaginary troubles of others is really delightful. He is absolutely sound in that, just as his complete indifference to the means of existence is beautiful, indeed magnificent. He will not make the smallest sacrifice of his intellectual inclinations—he will not play in public because of his contempt for the public, and because it irks him—although he plays divinely. I have never heard piano playing (except perhaps Liszt's) which gave me so much satisfaction—so light and clear, so cold and indifferent to passion."*

Clara can't deny that this is true. However, there is much one can forgive, must forgive, considering his father's rough ambitions, those childhood nights in the *Animierlokale*. He has confided in her after Robert has gone to bed, after the maid has banked the coals, the two of them lingering at the piano. She, in turn, has shared her own confidences, mostly about Robert's ill health which (she is convinced of this, *must* be convinced) is the result of persecution by those who refuse to understand that great minds cannot be bothered with all the little details of a mundane world.

Picture them there, priestess and protégé, Clara's work skirt pressing lightly, innocently, against Brahms's rough-stitched trouser seam as she speaks of her husband with a passion, a joy, she'd thought had been lost for good. Picture Brahms's gaze on her animated face, breathless at

---

* Bickley, *Letters from and to Joseph Joachim*, 129–30.

the thought of himself so close to a woman, this woman, who is untouchable in her genius, her motherhood, her position as Schumann's wife. Picture Robert upstairs, blanketed in a sleep so deep that for once he cannot hear his own mounting madness: devilish, dissonant notes that tormented him nightly, unceasingly, during the weeks before Brahms's arrival.

It is not impossible that all three of them might imagine things continuing exactly this way:

Clara's heart beating faster at the touch of Brahms's knee;

Brahms choked to shyness as her hand guides his own over secret fingerings;

Robert stroking those fresh, smooth cheeks, leaning forward to offer his kiss. And Brahms receiving that kiss from them both, Clara and Robert, Clara again. Returning it. Receiving it. Offering it again.

Morning walks together and lunchtime with the children. Afternoons spent at the piano, scratching at scores, debating each note with a passion that eclipses the page. Robert and Brahms composing variations on Clara's original themes, dueling at the piano, taking delight in Clara's delight as she turns effortlessly, gracefully, score to score. Hours of conversation in which everything and anything is talked about—

Except this thrum of longing that engulfs them. This warm sense of expectancy that sharpens every look, every word, into its own exquisite point. It cannot be identified because it is everywhere. They are breathing it. Like air.

Erotically, these types are torn in two directions . . . Where they love, they do not desire, and where they desire, they cannot love. The only defense against this dilemma consists in despising the object of one's sexual hunger while, conversely, adulating to excess the soul mate . . .

—Sigmund Freud*

---

* Harding, *Concerto*, 234.

25.

By six in the evening, Miriam and Chuck had closed up shop. No way would Hart be taking me up in the Blanik today. Nor was I going to be home by eight—though, somehow, this didn't seem to matter any more. Cumulus clouds compressed the horizon; at least the air had cooled enough to make it possible to sit outside. I packed up my computer and returned it to Hart's car, exchanged it for the short-story anthology, which I took to the edge of the pool. The Japanese students had retired to the dormitory; now and again, voices drifted through the door, which was propped open by a stone. For a moment, dangling my feet in the water, I thought I was hearing the piano.

Out of tune. Rasping. The G gone sour.

Sometimes it was hard to distinguish the world I'd reentered from the one I'd left behind.

I'd planned to look at a new Danticat story, but I found myself rereading, instead, "Where Are You Going, Where Have You Been?" by Joyce Carol Oates for what must have been the fiftieth time, seduced by the quiet foreboding that intensifies with each appearance of the stranger, Arnold Friend, who ultimately talks young Connie from the safety of her home, from behind her locked screen door, and into the car that will take her away. *The place where you came from ain't there any more, and where you had in mind to go is canceled out.*

Then I heard it again: the opening bars of Schumann's Arabeske. A pause, followed by another few bars.

Midori was practicing.

Puffs of wind rustled my pages; the weather vane shuddered, clocked around. Something was building: a system, a storm. I realized the sunlight was gone. At exactly the moment I started to worry, a hand rested cool against the back of my neck.

"I spent most of my adolescence in a swimming pool," Hart said. "Lap after lap in that cold, cold water."

"Do you still like to swim?"

He laughed. "I cannot bear the thought of it. Another lost passion, I suppose. Do you still play the piano?"

"I'm teaching my daughter."

"Ah." He settled himself beside me, unlaced his shoes. "But I have read this story," he said, peering at the page. "Lit-tle Connie meets the devil."

"You interpret Arnold Friend as the devil?"

"Don't you?"

"Yes and no."

He slipped his feet into the water beside mine. "I am thinking you are someone who fears commitment."

"And I'm thinking you are someone who believes in the devil, which is surprising, considering you don't believe in God."

"Arnold Friend is not *the* devil, he's *her* devil," Hart answered seamlessly. "Lit-tle Connie doesn't have a chance."

"Do you have a devil?"

"Everyone has a devil."

"Not me."

"You are certain of this?"

"Some people attract devils. Some don't."

"He is out there. Trust me."

I found myself thinking about the men—boys?—I dated during the years before my marriage. I thought about L—, his sincere and uncomplicated admiration for me. I thought about Cal, but it seemed impossible, now, that we could ever have aroused in each other anything beyond middle-aged weariness.

"No one would make me open that door."

"Someday, someone will coax you out."

"Fat chance."

His foot grazed mine in the water. "Me, perhaps."

"Don't flatter yourself."

He looked amused. "I thought not."

Another burst of Arabeske spilled over us both like a shower of gold.

"How many times have you been married?" I asked.

"Is it already time for true confessions?" From somewhere deep in a cloud, there was thunder. "I see that it is."

"Unless," I said, "you have something to hide."

He smiled. "Only my own stupidity. My first and third marriages were with Lauren. She was seventeen when we met. I was thirty-three." His mouth twisted ruefully. "An excellent age to meet your devil."

"Thirty-three or seventeen?"

"You are funny."

"How long were you together?"

"It is off and on. We married other people in between. The second time we divorced, you could say I did not take this so well. Friederike was ten, and I was the one who supervised her practicing, as you supervise your own daughter, yes? So you understand what this means. The time, the discipline. The commitment."

More thunder, sustained this time, followed by a sturdy gust of wind.

"The court had no such understanding. The child should be with her mother. *Basta.* Only now the child is old enough to have her own say. So, once again, as you have seen, the topic is under discussion." He tapped his shirt pocket, where he kept his phone, then looked up at the sky. "I'm afraid there will be no flying for you today."

"Too bad."

"You are truly disappointed?"

"I'd be lying if I said I wasn't relieved. But I'm starting to wonder if there's a connection between how nervous I am about—well, just about everything—and the way I've been stuck on my book."

"Maybe it would help you to fly, then?"

"That's what I've been thinking."

He glanced at me, surprised. Pleased. "I have always found it useful to try new things," he said, and then: "Help me with the glider, do you mind?"

A Chopin Polonaise accompanied us as we walked past the pilots' quarters. "Can I ask you another question?" I said.

"Sure, sure. The answers are free."

"Are you really a famous eye surgeon?"

This made him laugh. "Who is saying that?"

"Did you invent some kind of prosthetic eye that senses light?"

"My God, who do I look like? How do you say—the miracle man?"

"The miracle worker?"

"There is no such thing, Jeanette, as a prosthetic eye. Not the way you mean. None of this involves the eye, anyway. It is all about the brain."

But another gust of wind blew his words away, and by the time we'd rolled the ASW into the hangar, a dark wall of rain was sweeping toward us across the open field. Directly overhead, the sky lit up. The thunder was continuous, a hammering deep in my chest. Hart closed the hangar door, and when he turned to rest his fingers very lightly on my shoulders, I wondered if the regret in his face was a reflection of something he'd seen in my own. Twelve years of marriage. Richer or poorer.

The rough kiss of his chin. The spark of his tongue.

I'd believed we were forever, Cal and me. I'd believed only death could part us.

We kissed until the rain overtook us, until his hand found the small of my back, guiding me toward the trailer, toward the warm yellow light that shone from the propped-open door. The ax murderer. The entrepreneur. There was no need to hurry. We were already wet to the skin. We were already moving toward where this would take us, where this was going to end.

But for now, we'd eat dinner with the Japanese students. We'd listen to Midori perform Arabeske, and I'd play selections from Schumann's *Kinderszenen*, easy pieces that have lived in my fingers for over twenty years. We'd spend a restless night in the dormitory—the next night, too—Hart on the couch in the main room, me in a twin-size cot beside Midori. And it was on this second morning, the day of my first flight, that the sentence woke me out of my sleep, warm and glowing as the beam of sunshine splashed across my face.

That first flight was nothing like what I'd imaged. A few shuddering seconds on the tow, then the lovely, lifting feeling that means you're in the air. The sound of the wind pouring in through the vents. The creak of the rudder, like a seagoing ship. Hart released the towline at three thousand feet, and I found I was not afraid. Suspended by physics, surrounded by space, I looked out at the world as if for the first time: bitter-burned fields and gummy-eyed sinkholes, the single gray road that had brought me here. Everything familiar. Everything changed.

"Okay back there?" Hart called.

"Okay."

"You sure?"

But I was too happy to speak.

I made it home by seven that night, just minutes ahead of Cal. After that, I was busy with Heidi, who ran through the house touching everything—toys, furniture, even my face—as if to reassure herself that nothing had been lost during the time she'd been away. It took several hours before she was finally able to sleep. Before I could hurry down the hall toward my study, pinning back my hair. Eager as a woman going to her lover. That single sentence singing, still, its song inside my head.

What I wrote: *My first date in nineteen years is nearly an hour late.*

26.

WHAT MAKES YOU PASSIONATE *about flying?*

Character come before person is born.

What if something happened so you couldn't fly anymore?

I am careful. Nothing happens. You will see!

What do you think makes some people want to do what you're doing and other people, like me, afraid to do it?

Do not eat uncook vegetable. Begin each day with hot boiled egg.

Part V

✖

Translation

Düsseldorf, 2006

27.

It's the first trip they've taken together, aside from short jaunts to visit friends. Over five days' time, they'll hike over one hundred miles along the Rhine. They'll climb the Lorelei. They'll picnic in the shadow of crumbling castles, explore the little villages scattered through the hills. Sweet stone churches. Half-timbered houses. Johannes talking and talking and talking, bubbling on like a fresh, clear stream. Clara listens to his voice the way she listens to music, breathing hard from the climb.

Passing the shared canteen.

Something about the distant clearings of steeply sloped vineyards and gable-chinned roofs, woods and more woods of linden and oak, makes her feel as if anything is possible. It's been sixteen months since Robert entered the asylum, but she doesn't think of this. She doesn't think of the children. She doesn't consider what people might say. For once, she can taste the dark scent of the Rhine, far below, without weighing the question of whether or not it might have been better had Robert—

Now and again they stumble upon an ancient orchard, an overrun garden, a half-concealed well greenly tangled with ivy. Once she turns to find he has vanished. But no, he's scrambled up into a mulberry tree. Laughing, he pelts her with fruit. They walk on. Touch each other swiftly, in places that do not matter. Here he is combing bits of fern from her hair; there she straightens his rucksack, kicks pebbles at his shoes. When they stop to rest, she eats the crushed mulberries he offers, still warm, from his pockets.

The news about Robert isn't good.

At some point, she must consider propriety. What will be the consequences of such a trip? It is one thing, while she's on tour, having Johannes constantly in her home, busying himself with the children, keeping an eye on the servants, documenting household income and expenses in his quick, vigorous hand. Especially since his devotion to her husband is so well known. Still, it would be better, upon their return, if Johannes would agree not to nap on the couch, lazy and warm as a loose-limbed cat, where visitors can see him plainly. If he wouldn't write so admiringly of her to mutual friends. If he'd stop fussing over her like a husband each time she left on tour, urging her to pace herself, to take care of herself, to come home soon. If he didn't insist they address one another not as *Sie* but *Du*. Not only in letters, but in company, trumpeting an intimacy that would strike anyone as peculiar. Not that there's anything between them that isn't open and good.

They are friends. *Best friends*. What else can they be?

In this light, after much discussion, she has finally agreed to return his warm and affectionate *Du*.

There are, of course, his dark moods. Understandable,

considering his childhood, his past. Easily forgotten—or, at least, explained away—as soon as they have vanished, which they do. And then, how easy, how effortless everything is! The meals they share, the walks they take. Conversations about the children: what is best for the girls, what should be done with the boys. Discussions about Robert's situation in Bonn, doctors' reports, new therapies. Analysis of Clara's performances, program selections, resulting reviews. Debates over Johannes's compositions, which seem, at times, to belong to them both.* Together, they sit at the piano, reviewing modulations, refining passages, Clara's fingers moving over the keys like gestures accompanying a story she's determined to finish. It is not enough that he says he understands.

She bats at his arm. *Do you see?*

In the end, of course, he listens—as she listens to him, in her turn. He encourages her own compositions, applauds her private performances, yielding to her superior sense of line, her inspired fingerings. He distracts the children with gymnastics, walks, giving her the gift of a clear afternoon in which she can sit down to compose. He intercedes on her behalf with doctors and bill collectors, and when a rat scuttles out of the coal bin, it's Johannes who brandishes the broom, screaming along with the children, even as he chases it through the parlor and out the front door.

And then there's the way he talks to her, of everything, of

* "By rights," Brahms would write to Clara in 1891, "I should have to inscribe all my best melodies, 'Really by Clara Schumann'" (Reich, *Clara Schumann*, 202).

nothing. Of music and sunlight, of trees and stones, of porridge and history and science and God. Of his mother. Of her father. Of a restaurant in Dresden where, once, he sampled a particular kind of cheese. Of local politics and musical fools, America, the plague, French fashion, Liszt, Fanny Mendelssohn's new parlor maid, Joseph Joachim's annoying tendencies, German folklore, Hamburg. She's following him down a steeply sloped path: thirty-six years old; strong-willed; strong-limbed. But she cannot keep up with all that he says. She cannot keep up with her own flooding heart. Placing her foot against a loose stone, she tumbles forward as he turns to catch her, touches her forehead to the rough kiss of his chin. Somehow he's holding her hard in his arms, his exhalations filling her inhalations, an exacting completion, a jigsaw fit, and she thinks of making music. She thinks of making love. The places on her body he has not yet touched are dark spaces.

The rest of her shines.

Do I want to have Robert back like this? she has written in her diary. *And yet, should I not want to have, most of all, the person back again? Oh, I don't know what to think anymore: I have thought it all over thousands upon thousands of times and it always remains just terrible.**

Wake up, she tells herself. Step away.

He keeps breathing. She keeps breathing. Around them, the green land. The golden light of dreams.

This is not the sort of man with whom you build a future.

At some point, you must step away.

---

* Reich, *Clara Schumann,* 150.

Date: Saturday, June 17 10:58 PM

To: LMJPROF@que.edu

Dear L—,

You wanted to know how I really am. I needed some time to think about my answer.

Six weeks ago, I met a German man—actually, I can't shake this déjà vu feeling he's someone I've met before—and I suppose we're friends, though he insists that men and women can't be friends. He also says there's no chemistry between us, but lately it seems there's some chemistry after all. Remember the book I always wanted to write about Clara Schumann and Brahms? He's helping me with translations. Also, I'm going to Germany next month, and he'll be there at the same time. Sometimes I think he cares about me. Other times, there's this calculating distance, as if I don't quite measure up.

So all this is to say that I guess I don't really know how I am, aside from being sorry, from my heart, to hear about you and Sally. Is it something that can be fixed? Then again, I suppose these things can't ever really be fixed. But maybe it's different with a second marriage. Maybe things are easier to figure out.

Heard any more divorce jokes lately? I've started collecting them. It helps. Anyway, I'm thinking about you.

Jeanie

Date: Saturday, June 17 11:16 PM

To: Jeanie88@comster.com

Hi Jeanie—

How about this one? Half of all marriages end in divorce, but the other half end in death. Sally told me that over lunch yesterday. We've agreed to divide things up nicely. We're putting away our knives. I suppose that's the biggest difference between a first and second marriage. With the second, you're much more willing to cut your losses when you see that things are wrong.

About this guy who's helping you with the Clara book: if men and women can't be friends, what, exactly, are the two of you—research associates? Research associates with benefits? Look, either there's chemistry or there isn't. And that déjà vu feeling like you've met him before? It crosses my mind—now don't get mad, Jeanie, but you did talk to me, once, pretty frankly about Cal—that you're setting yourself up yet again to want somebody who won't want you. There are plenty of men who find you attractive, and I don't just mean your beautiful mind. What if you hung around with one of those guys for a change? Especially since you're a free agent now. You wouldn't have to leave him outside your door this time.

When, exactly, do you leave for Germany? Did you get a direct flight or are you coming through New York? If you're still speaking to me, I could meet you for a drink.

I hope you're still speaking to me.

L—

28.

WE MET IN SUSHI bars, at coffee shops, at open-air restaurants fronting the beach, lugging backpacks and satchels, overstuffed folders, Internet printouts bunched together with clips. I'd leave Heidi at my parents' house for dinner, then head out to meet (so I explained) a German man who was helping me with research on my book.

"How did you find him?" my father asked.

"Through a service," I said truthfully.

"Is he charging you money?" my mother wanted to know.

"Spell everything out up front," my father said. "You don't want trouble down the road if this guy wants royalties or something."

This particular night at the end of June, we were at Crabby Joe's overlooking the boardwalk, and even before the specials had been read, our table was covered with papers and Post-its, biographies and articles, a scattering of pens. The waitress returned again and again, but no, we were not ready, we needed more time. We were talking about the first vacation Clara and Brahms took together, in 1855, a time which Clara described in words like *rapturous* and *exquisite.* We were talking about the second vacation they took together in 1856, two weeks after Robert's death: the ill-fated trip to Gersau. We were talking about the letter Clara sent to Brahms in 1858, in which she writes, *If only I could find longing as sweet as you do.* We were talking about the letters between Brahms and Clara in the 1860s, in which Brahms declares himself an "outsider" to Clara and her children. By then Clara had purchased a house in Baden-Baden, where she'd vacation with the children each summer. Frequently, Brahms joined them despite moody outbursts that left Clara in tears.

"There was a small demon in him," Clara's youngest daughter, Eugenie, would recall, "and who does not know from experience that we are apt to let it off against those of whose affection we can always be sure? This was the case with him. What Brahms loved in our mother above everything, above even her artistic understanding, was her great heart; he could be sure of its love and forgiveness even if he were to let loose with a legion of demons."*

"He wants to be close, he can't stand to be close," Hart said, closing his edition of *Erinnerungen.* "I tell you, Jeanette, there are things about men and women that do not change. She welcomes him in, he shoves her away. She kicks him out, and then he wants to be close again—"

---

* Schumann, *The Schumanns,* 157.

"But *why* does he want to be close again? She had a great heart, but as long as we're generalizing here, isn't that what women are known for? There were plenty of great hearts for Brahms to torment. Women were always falling in love with him. I'm thinking, despite what Eugenie says, that what brought him back to Clara—what kept them close even after things didn't work out in a romantic sense—was exactly that artistic understanding—"

"Artistic understanding has nothing to do with it! Jeanette, I promise you, this same story could be told about two butchers, about two field laborers—"

"A field laborer can talk to other field laborers. But who could either Clara or Brahms have talked to at this level of expertise—"

"Mendelssohn, Wagner, Liszt—"

"They hated Liszt—"

"What about your ex-husband?"

"What on earth are you talking about?"

"The other night, after he brought your daughter back? When he upset you so?"

"I shouldn't have told you—"

"You told me he said he was feeling *like an outsider.* Didn't he use exactly those words? So you felt bad for him. You invited him in, but as soon as he is sitting on your couch, he . . . how do you say . . . lashes out? So, of course, you get very upset while he, on the other hand, feels better for a while. But then this wears off. He feels sorry for himself. He feels sorry for you. He returns to your door. Once again you ask him inside—"

"That's different—"

"Just *stop.*"

"Okay, but we've only been divorced ten months—"

"And how many times has this happened before? How many times will this happen again—"

"Seriously, Hart, I think we'll be friends. I want us to be—"

"Because of your shared musical genius, ah! Your artistic understanding! Yes,

I see everything clearly."

"My God! That's *so* unfair, so cynical, I—"

"You want unfair and cynical? Look here."

Hart passed me an airmail parcel. Inside was a copy of *Aus dem Kreise Wieck-Schumann,*[*] a slender, bitter recollection penned by Clara's half sister, Marie.

Instantly I forgave him everything. "Where did you get this?"

"Friederike sent it to me. Thursday, I must catch the red-eye to L.A. I'll take a look at it then."

"What's in L.A.?"

"A conference I must attend." He reached for my journal. "May I?"

"A conference on what?" I said, but he was writing notes beside my own in dark, meticulous pen. I thought of Friedrich Wieck writing in Clara's diary, composing first-person entries as if his voice and his daughter's were one and the same.

I'd never allowed anyone to read my journals.

Much less write in them.

After dinner, we walked along the boardwalk. The sun was still high, but the air remained humid, thick. A sour smell rose from the beach. Hart was talking about the fuel crisis, about the Bush administration's isolationism, about sea turtles. He was talking about a woman he'd dated who

---

* *From the Wieck-Schumann Inner Circle.*

disliked wearing shoes. He was talking about the weekend weather forecast, which didn't look good in Orlando, but was promising south of Miami. He belonged to a private glider club there—maybe I'd like to come with him? He owned shares in an ASK 21 there, a two-seater, much nicer than the Blanik.

"How did you get your name?" I asked, interrupting the torrent of words.

"I am named for a grandfather," Hart said. "And you?"

"But *Hart*?" I said, refusing to be distracted. "Is that the usual nickname for Reinhardt?"

"Lauren started calling me that." A gambling boat was making its way out to sea; we stopped to watch its progress.

"Because of how much she loved you?"

"Because she thinks I have none." Hart bent to touch my forehead with his own. "Where does it go, the good feeling between us?"

"Between you and me?" I said, surprised.

"You and me, sure. You and your Cal. Me and Lauren."

"It doesn't have to go anywhere."

"Ah, Jeanette. It is always the same thing."

"It wasn't the same thing for Clara and Brahms."

I tried to put my arms around him, but he sidestepped me awkwardly, angrily, hopping down off the boardwalk and onto the sand. "You have done a lot of research on your Clara," he said. "As a woman, perhaps, you understand her better than I. But I am a man and I can assure you, what happened between them is no mystery. Listen to how Brahms writes to her: *Do not ever throw away a* *pretty ribbon from your hat, or anything of that sort, but give it to me so I may tie it around your letters.** How long could any man, especially at that age, be satisfied with letters and ribbons?"

"Well, if all the whorehouses in Hamburg couldn't satisfy him either—"

"Let me put this to you plainly. He's young enough to think that Clara will be different. After Robert dies, they travel to Gersau, where he fucks her and discovers she is not."

He stalked away toward the water. I started to follow, changed my mind, headed up the beach toward the pier. The water's edge was lined with dark ribbons of seaweed, the cracked ruins of turtle eggs, trash. Ugly landscape. Ugly words. I was tired of this. Tired of him. He could touch me, yes, but on his own terms. He could say anything he wanted. *He wants to be close. He can't bear to be close.* This was going nowhere. When I got back to the car, he was waiting for me, leaning against the hood. We stared at each other, unblinking.

If only I could find longing as sweet as you do.

"What are your feelings for me?" I said.

"Don't ask me that," he said. "Please."

At home, I opened my journal, placed my hand where his hand had been. Traced his careful writing with my finger.

Then I tore out those pages. Took a long shower. Washed the smell of his cologne from my hair.

---

* Litzmann, *Briefe*, 185.

Permit a poor outsider to tell you today that he thinks of you with the same respect he always did and from the bottom of his heart wishes you everything good, much love, and a long life to you, the dearest of all persons to him. Unfortunately, I am more of an outsider to you than anything else.

—Brahms, in a letter to Clara, 1892*

---

* Reich, *Clara Schumann,* 198.

29.

We'd put Heidi down together, taking turns reading to her from *The Sleep Book*, the way we'd almost never done when we were still living under the same roof, planning, step by painful step, to break into pieces everything we'd spent years building together. After she'd fallen asleep, Cal stood in the kitchen, talking about his students, while I loaded the dishwasher, wiped the countertop, turned off the kitchen light.

Still, he didn't go. It was a long drive back to Lakeland. "I feel like such an outsider," he said.

"Want something to drink?"

"Coffee would be great."

I'd expected he'd ask for wine. But he didn't look as if he'd been drinking much these days. Everything about him seemed cleaner, calmer, and I wondered how much of what he'd suffered—the anger, the depression—had been tied to me, the way that my own fearfulness, my anxiety, had been tied to him. His skin was clear, no puffiness under the eyes. He looked like a much younger man. I, too, was looking younger. Healthier. I'd bought new clothes for the first time in years. I wore earrings. I'd recently colored my hair.

"Oh, I meant to tell you," he said. "I finalized vacation plans for Heidi and me."

"What are you going to do?"

"Dad's summer place the first week. Dan and Kelly are coming, too, with the kids. After that, I'm taking her to a reenactment in North Carolina. We'll camp out, if she's into it, but a friend of mine's renting a room at a hotel, so we can always crash there, if we need to."

"Perfect."

In my living room, formerly our living room, Cal sat on my couch, formerly our couch, while I made coffee the way he liked it: boiled on the stove, thick with cream.

"You're seeing somebody, aren't you?" he said as I carried in the steaming cups.

"Why do you say that?"

"You're ready to take off as soon as I pick her up. Then you're late coming back."

"I was late tonight," I said, trying hard not to sound defensive, "because I was working with someone who's helping me with translations for the book. Twenty minutes. And I called to let you know I was running behind."

Still, it was twenty minutes he and Heidi sat in the driveway, Heidi staring at the locked front door of her own house. *Offer him a key,* I thought. The words were on the tip of my tongue.

"Why don't you just admit it? You're seeing this guy, right?"

"I didn't even say he was a guy."

"Why can't you be honest with me? At least we were always honest with each other."

"We were always honest," I agreed.

"Well, I'm seeing someone," Cal said, picking up his coffee. "She'll be at the reenactment. She's the one renting the hotel room. Oh. You put in cream."

"You don't take cream anymore?"

"I watch what I eat these days."

"You look good," I said truthfully.

"So what's he like, this translator? The one you're not seeing?"

I shrugged.

"Rich, I suppose. Probably an attorney. Everything your father always wanted." His tone perfectly poised on the line between joke and accusation.

"He isn't an attorney. He likes to read. He's interested in classical music."

I was sounding like America now, and Cal deserved something better, didn't he? I struggled to find something honest to say. Something generous and healing.

"Sometimes he reminds me of you," I said. "The way he likes to talk about things. The way he likes to debate." *The way he holds himself aloof.* In a moment, I'd say that, too. So I said, "But what about you? Your girlfriend, I mean. Tell me what she's like."

Even though I was looking right at him, I never saw it coming. Cal put down his empty cup.

"She is absolutely nothing like you. Do you think I would make the same mistake twice?"

There are things between men and women that do not change.

Of course they were lovers, Clara and Brahms. How could I ever have thought otherwise? Who else but a lover retains the ability to wound the other person with such passion, such precision? And who else but that lover has the capacity to heal what he or she has done?

I only want to ask you not to transform people into an enthusiasm, through your own, which they will afterwards not understand. You demand too fast and impassioned an acceptance of the talent that you cherish. Art is a republic. You should make that more your principle. You are much too aristocratic . . . Do not assign a high rank to one artist and demand that smaller ones should regard him as superior . . . Do not consider my folk songs [translator's note: it seems he has enclosed them] as more than the most sketchy studies . . .

—Brahms, in a letter to Clara, 1858*

Dear Johannes, of course you do not see or hear that when I talk about you with others, I truly do not do it in exaltation. Yet that I am often mightily gripped by your rich genius, that you always appear to me as someone upon whom the heavens rain their most beautiful gifts, that I love and adore you for so many glorious things—that this has taken deep root in my soul, this is true, dearest Johannes. Do not try to extinguish this within me by cold philosophizing—it is impossible . . . Why should you wish, by your coldness, to kill the beautiful confidence which allows me to tell you anything? You have already done so, for regarding your folk songs, I am afraid to tell you the happiness most of them have given me . . .

—Clara, in a letter to Brahms, 1858†

I recently read something about enthusiasm in a letter that Goethe wrote to Schiller . . . where he is saying with a certain reservation, carefulness, etc.: "I always have the feeling that

---

* Litzmann, *Briefe*, 222.
† Ibid., 223–24.

when writings and deeds are not talked about with loving concern, with a certain biased enthusiasm, little is left of them, they are not worth mentioning . . . Pleasure, happiness, and partiality are the only truths that bring forth reality." When Goethe is saying this, should I not feel above your reprimand?

—Clara, in a letter to Brahms, 1858*

I am sorry I did not write to you about the Hungarian dances [author's note: which Brahms sent to her], for you know how I like to please you. I only refrained because I feared that you might say something unkind to me, as you have often done in similar cases before. You know, without my telling you, how hard it must have been for me, because it would have given me the greatest joy to write to you about them . . .

—Clara, in a letter to Brahms, 1858†

When I first got the news of the unfortunate reception of your concerto I straightway sat down to write to you. I felt that a kind word would be a solace to you. But then I was afraid that you would answer me shortly and that I should feel offended. . . . Did you not try the Serenade at all? If you had played this first, your victory would have been certain, because it is a much clearer work.

—Clara, in a letter to Brahms, 1859‡

---

* Ibid., 227.
† Litzmann, *Letters*, 89.
‡ Ibid., 96.

30.

One week passed before Hart called again. "We had talked about flying the ASK," he said. "If you are still interested."

"Hello to you, too."

"You are angry."

"You are complicated."

"I missed you."

"I missed you, too."

The private club consisted of three gliders packed into a small, flat-roofed shed, a dented Porta Potti, and a weather-warped picnic table protected by a sun-beaten canvas. We had to call in a tow pilot from the county airport, two miles away, but the lift was good and we were up for two hours, gliding twenty-five hundred feet above the Everglades, cutting from cloud to cloud until we reached the coast. Yellow ocher water hugged the beach, then darkened as it deepened, enriched with cobalt blue. There were white-ribboned breakers, luffing sails. The dissipating wake of a speedboat dividing around a hard-knuckled reef. Sand sharks migrating along the shoreline, finger-shaped shadows just a few feet beyond splashing tourists; kids on surfboards; a black, paddling dog. We began to thermal, making tight, dizzying circles within a warm column of air. Soon we were joined by black-headed vultures, open-winged, rising beside us. But by the time we reached five thousand feet, it was nearly five o'clock.

"Better head back," Hart said.

"Already?" I said.

"Unless you want to land out on the beach?"

"No, thanks."

He'd been uncharacteristically quiet all day. Or perhaps I was the one lost in thought, looking down at the drained and drought-stricken land, such a contrast from the ocean's motion and light. Two weeks from now I'd be in Germany, and when I thought about all I had to get done before I went, I wanted to put my head in my hands and weep. Other than attending Friederike's concert, I hadn't made any firm plans, and even the concert felt tenuous, connected as it was to Hart. Who knew if we'd even be speaking once another two weeks had passed? I wanted to see the Robert Schumann house in Zwickau, but I hadn't figured out how I'd get there. I had a vague idea about visiting Bonn and Düsseldorf. Maybe, if there was time, heading south to Gersau. Poking around Lake Lucerne. Staring up at the Rigi.

"Tell me again when you're leaving for Europe," Hart called back to me.

I was no longer surprised by how frequently he responded to my thoughts as if I'd just spoken. There were, I supposed, explanations.

"The fourteenth. A few days after you."

"You can stay with me in Leipzig if you're interested. I have been offered the apartment of a friend."

I stared at the back of his head. "I don't know."

"What don't you know?"

"If this is a good idea."

"You need a place to stay. I have a place to stay. Now who is being complicated?"

Back on the ground, I helped him roll the ASK—lighter than the Blanik—into the shed, and then we sat on top of the picnic table, drinking Gatorade as the sun shaped itself into a molten ball above the horizon. The fields around us were planted in squash, but the crop was small, wizened. Here and there, tattered blossoms caught at the light, flashes that reminded me of fireflies. A killdeer called, fluttering, in the dust.

"I will meet you at the apartment," Hart said, as if there'd been no break in our conversation. "There are trains from the airport; it is easy. You can have a little nap, if you need one, and then we'll take the tram to Ingelstrasse. Maybe stop for a bite to eat first, if there's time."

"What if I just meet you at the concert?" I suggested. "You'll be spending the day with Friederike, right? She won't want to share you with a stranger."

"In some ways, I am equally a stranger." Hart leaned forward, pinched a mosquito from my arm.

"Come on, you talk to her all the time. Besides, you just saw her in London."

"For the first time in over six years, yes."

I thought I'd misheard him. "What did you say?"

"It is true." He wasn't looking at me. "The last time I saw her, she was ten years old."

The sun slipped closer to the horizon's edge. Flat strips of clouds haloed Miami, a tapestry the color of industrial waste: neon purple, electric orange, lavender, mottled green. I tried to imagine not seeing Heidi for one year. For one month. The two weeks I'd be in Germany without her already seemed like an unspeakable loss.

"You will want to visit Zwickau, I imagine," Hart continued. "Perhaps Düsseldorf and Bonn. Friederike is interested in this as well, so I'm thinking it is best to rent a car. At the end of the week, she goes to music camp in Zurich. If her mother agrees, we'll drop her off there, and then you and I can go on to Lucerne, take a ferry to Gersau. There's a glider club nearby as well. We could fly a little way into the Alps."

The warmth of his shoulder had found my own. I leaned back against it, expecting he'd pull away. But he put his arm around me like a man claiming a decision, held me awkwardly, determinedly, as the sun dropped below the edge of the field. Together we stared at the poisoned sky, listening to the sounds of night insects and, in the distance, I-95. I felt as if I were being embraced by—not a stranger, exactly—but someone I knew very slightly: a bank teller, a crossing guard, a checkout clerk at the grocery store.

"What is it you're not telling me?" I asked.

Suddenly the air was clammy, cool, the way skin feels after fever. I hadn't even realized I was shivering until Hart pulled a rumpled shirt from his flight bag, draped it loosely, kindly, around me. With that gesture, he became himself again. His arm around me lightened.

"Certain things," he said. "As, I suppose, there are things you do not tell me."

31.

"He's married," Ellen said. "Or gay."

We were sitting beside her backyard pool, where her twelve-year-old niece, Mabel, had been entertaining Heidi for the past two hours: giving her underwater rides, teaching her to cannonball, spinning her in circles on an inner tube.

"Nope," I said, reaching for my lemonade. "And nope."

"How often do you see this guy?"

"Depends on . . ." I nodded at Heidi.

"How often do you talk on the phone?"

"Every day. Your point?"

"But you're not dating. It's not serious."

"He's helping me with the book."

"*Your* book. Why is he being *so* helpful? And when does he have time to do all this reading? Maybe he's not really a doctor. For all you know, he doesn't even work."

"He had a stack of mail in his car one time, all of it addressed to Dr. Hempel. And once, when I called him about something, he said he was at the lab."

"What lab? Where?"

"He said he couldn't talk because he was balancing a stack of slides."

"Slides of what?"

"Monkey brains."

"Really?"

"How should I know? Look, Ellen, it just doesn't matter."

"Of course it matters. *You* matter. What are his feelings for you?"

"Murky," I said. "Is that a crime?" But my eyes filled with tears, and I hated myself for this, hated myself for being drawn into a conversation I didn't want to have. "I know you think he's using me, but I'm writing about all this, so you could say I'm using him, too. Or maybe there's a kinder way to look at it. Maybe we're helping each other somehow."

"Sweetie," Ellen said. She put one cool, ringed hand to my cheek. We waited, and after a moment I said, "I've seen where he lives, does that reassure you? Beachfront condo. Direct ocean view."

"From the master bedroom, too?"

I gave her a look that made both of us laugh, so it was okay again, even though it wasn't. "I wouldn't know. We just stopped by to grab a book. But from the looks of the living room, Viso-Tech does just fine."

"So then why won't he talk about it? And what's this big secret with his daughter?" The girls had clambered up out of the pool, breathless and wrinkled and dripping. "Can we have something to eat?" Mabel asked, leading Heidi across the patio toward the cabana. Ellen nodded, lowering her voice. "C'mon, he doesn't see her for *six* years? Maybe he A-B-U-S-E-D her or something."

"Please."

"You're always with this guy. I hardly ever see you anymore." She poured herself more lemonade, her pretty face flushed, brooding. It was true that I hadn't been home much lately, but it was also true that when I did call, she didn't call back either. She was spending lots of time doing volunteer work at a women's shelter. She'd started jogging, evenings, with her sister. As far as I knew, she hadn't gone out with anybody since Dancing Man. "Before you go, I want you to write down his full name and the name of his company. I'm going to plug him into our system at the bank, see what I can find."

"Ellen."

"Meanwhile, when you *do* sleep with this guy, use an industrial-strength condom."

I glanced at the cabana. The girls were scooping ice cream from the well-stocked fridge into colorful plastic bowls. "I'm not going to sleep with him," I mumbled.

"You said he was cute."

"I said he was handsome."

"Why not, then? If the situation arises." She laughed bitterly. "And trust me, it will. It always does."

What I said: Absolutely nothing. Because what was there to say?

The master bedroom also overlooked the ocean. The bed, king-size, was checkerboarded with books: fat medical texts, novels, trade journals, biographies. Before he slept, he read for hours, devouring information, remembering every word. I could have told him, then, what I'd been writing. But I didn't. The one thing he didn't want to know was my heart.

"I envy you your passion," he'd said, the first time, the sliding doors open to the sounds of the sea. "Once, I loved my research, I think, the way you are loving your writing. Now I am a dead man, I must warn you."

"You don't seem dead to me."

"You mustn't get attached to me, Jeanette. I am a dead man. I am like a stone."

32.

Just before he left for Germany, Hart came to my house for dinner. It was the first time he'd entered my other life, my real life, the life in which I worked and wrote, visited my parents, raised my daughter. He set the table and tried to chat with Heidi, who regarded him suspiciously, and with uncharacteristic reserve. He ate, without comment, the overcooked chicken and green beans. He looked strangely ordinary. Small. Afterward, he cleared the table and washed the pans while I gave Heidi her bath. "When is that man going home?" she whispered as I tucked her into bed. When I tried to step away, she gripped my arm with a strength that startled me. "Don't go," she said, tears spilling sideways into her ears.

Half an hour later, I came out of her room to find Hart at the piano, paging through her music. "She's already playing Bach?" he said.

"Just starting. Yes."

"She is cute."

"Thanks."

"She looks nothing like you."

"Thanks again."

He asked where I wrote, and when I took him down the hall to my study, he looked up at my treasured Gaela Erwin self-portrait and let out a little mock shriek. "A man must have a drink after such a shock," he said, settling himself on the love seat, and before I could tell him that I usually—okay, never—let people into my work space, we were sitting together, drinking wine, select chunks of the manuscript spread between us as I hunted for places where I needed a line in German or French:

Kann ich denn nicht mit dir kommen?

Il faut que je te laisse.

And then we were talking about the Schumann children: Emil, who died in infancy; Julie and Felix, who succumbed to tuberculosis; Ludwig, who was committed to a mental institution where he died, like his father, of undiagnosed causes. Ferdinand became addicted to morphine, leaving Clara to support his wife and seven children. Clara's two oldest daughters, Marie and Elise, along with their youngest sister, Eugenie, fared better, at least in physical ways, but it was to Marie and Marie alone that Clara would turn, throughout her lifetime, as both daughter and friend. The other children were provided for in every practical sense, but they were raised by servants, boarding school teachers, and eventually—the youngest, at least—by Marie. Clara saw them on holidays plus a few weeks each summer, if then.

"Not an uncommon upbringing for children of that class, in that place and time," Hart said. "Once again, you are finding *significance*"—he squeezed my foot—"where there is none. I am thinking this longing for significance is a serious flaw of character."

"Which would, in itself, become significant over time. Think of all the decisions such a longing would influence. Think of all the ways it might alter, significantly, the direction of my life."

"Ah, Jeanette."

"Ah, yourself. You don't think it's significant that Clara concealed Julie's death so that she could go ahead with a scheduled concert? That she only visited Ludwig once during the many years he lived in the mental institution? That Marie, not Clara, attended Felix as he lay dying?"

"Distinctive, perhaps. But significant?" His hand was on my ankle now. The back of my knee. "In a thousand years, what can such a word mean?"

"I don't plan to live a thousand years. Do you know how Clara informed her two oldest children of Robert's death? In a letter. Do you know how often she saw Ferdinand during the year before his death? Not once. Marie went to visit him in the hospital. Marie, for that matter, arranged his funeral. Meanwhile, Clara's writing in her diary, *Work is always the best diversion from pain.*"*

"Sex is also effective," Hart said, shifting his weight. Manuscript pages fell to the floor; somehow he'd pinned my arms. "Is your daughter a good sleeper?"

"No. And then I read something . . . can't think where . . ."

---

* Reich, *Clara Schumann*, 169.

"Might we lock that?" He was eyeing the open door.

" . . . about Clara traveling to perform in some major German city and encountering on the street, *completely* by accident, one of her daughters who went to boarding school there. A daughter she hadn't seen in months. The two of them spent *a most delightful afternoon* before it was time for Clara to return to her hotel."

Hart sighed, rested his head on my chest. "You are thinking it is cold, even heartless, that Clara wouldn't have made plans to see this girl."

"Yes and no."

"I am tired of the yes and no."

"I just mean that I understand how she felt. You think I don't miss the uninterrupted intellectual life I had before Heidi was born? But I wouldn't put her in a boarding school, even a good one, for the sake of that work."

"No, you'd just put her to bed at your parents' house. You'd hire a babysitter. And don't tell me you are not happy to have her father take her for two weeks so that you are free to travel—"

He was right, and it cut me to the quick.

"At least I wouldn't miss her birthdays! I wouldn't skip her graduations! I certainly wouldn't leave her to die alone in a hospital or mental institution because it interfered with my writing. There's a balance to all this, and I'm not saying I've got it right—I *know* I don't have it right—and I also know, by the way, men do this sort of thing all the time—"

"Men like me, is what you are thinking."

"Which is why you've been successful at the things you've pursued. Maybe I'm just jealous she was able to do it, too. I suppose I believe, in the back of my mind, that if I were a true artist, a real artist, I could."

His weight deepened against me, the way a child's weight deepens with sleep. "You are a strong person, Jeanette. You would have visited your husband in the madhouse, I have no doubt of this. You would have sat at the bedside of each and every one of your dying children. But not all people are strong in this way, particularly when there's a child concerned."

"Does it take so much strength to let your lonely little daughter know you'll be in town for a day?" I was speaking against the side of his neck; he could not see my face. But he knew—he must have known—what I was thinking.

Or visit your daughter more than once in six years?

His breathing matched my own.

"The first weeks, the first months, after you are leaving a child?" he said. "It is a difficult thing. Not knowing how much she is changing. Having no connection to her daily life. Nothing to talk about."

He was too heavy now, but I didn't want to shift his weight, didn't want to risk distracting him, interrupting whatever it was that, at last, he was going to say.

"Eventually you start feeling better. You hope it goes the same way for the child. You don't want to stir it all up again, for either of you, so you stay away. You send the checks. You make sure she wants for nothing. Do you understand what I'm saying? I fought for my daughter in court, and you know what they gave me in the end? One weekend each month. One *supervised* weekend. And me living overseas."

I wriggled a bit, I couldn't help it, and he said, "I am crushing you, I think."

"No," I said, but he pushed himself away from me, the excuse he was looking for, and then we were sitting in our separate corners, as far from each other as, just moments earlier, we'd been close. He touched his wineglass, tapped the base of the lamp. He picked up the plastic hair band that had fallen out of my hair and fiddled with it, twisted it, until it broke.

"Sorry," he said. Then he broke it again. I got up and locked the door. When I turned, he was behind me, he was kissing me too hard, as if he were testing me to see what I could bear. At that moment I was terrified. All I wanted was to return to the relationship we'd had before. Because one can write anything about a dead man.

One has no obligation to a stone.

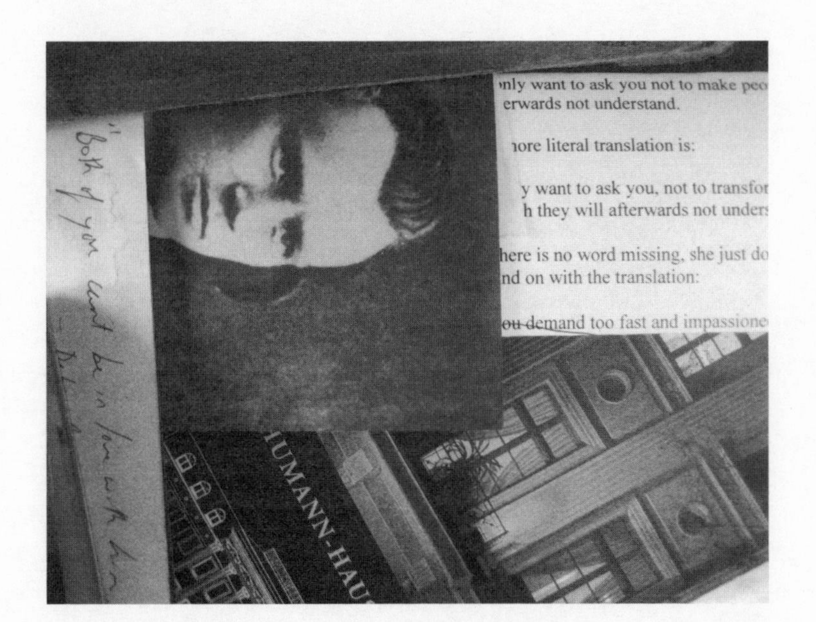

only want to ask you not to make peo
erwards not understand.
nore literal translation is:
y want to ask you, not to transfor
h they will afterwards not unders
here is no word missing, she just do
nd on with the translation:
ou demand too fast and impassione
UMANN-HAU

33.

I WAS LATE GETTING up in the morning, rushing to wake Heidi and help her dress, fix breakfast, get her lunch packed for school. She sat at the counter, picking at her toast, and I noticed her face looked flushed. "Do you feel okay?" I asked, placing my hand on her cool forehead.

She shook her head. "I had a bad dream."

I was relieved; she could still go to school. "What about?" I said, reaching for half of her toast. "Come on, I'll eat this piece. You eat the other."

"A man took you away."

"Honey," I said. She was staring at her plate. I pulled the other counter stool close and sat down. "It was just a dream, okay?" I told her. "Everyone has dreams. What did he look like?"

"Little." She held up her thumb and index finger. "Like this."

"Tell you what," I said, trying to make her smile. "If I see him, I'll step on him. I'll smack him with a rolled-up newspaper."

As if released from the spell of her dream, Heidi picked

up her toast and bit it. "It won't make any difference, Mom. There's nothing you can do."

Suddenly I felt disoriented, as if I were the one in a dream. For the voice I heard, though Heidi's, was the voice of someone older. The girl she was becoming. The woman she would become.

Everyone has a devil. My Heidi would be no different. When he came for her, she, too, would recognize him. She'd open that screen door.

Date: Wednesday, July 5 10:32 PM

To: LMJPROF@que.edu

Dear L—

Sorry for the silence. I wasn't offended, just busy. And thinking, too, about what you said. About history repeating itself. About everything. So you really think I have a beautiful mind? I've always liked yours, too.

I'd love to meet you for a drink—I'm connecting through JFK—but I have to tell you I find myself really involved with this man I told you about. Everything you wrote to me makes sense, and yet I don't want to back away from what I feel. The way I would have done in the past, as you of all people should know. Perhaps this is why you're the one person I really want to talk to now. If that's unfair, or unkind, you will tell me. You will always give me an honest answer, even when I don't want to hear.

xx,

Jeanie

Date: Wednesday, July 5 10:35 PM

To: Jeanie88@comster.com

Hi Jeanie,

I am always happy to see you. That's my honest answer, though it's incomplete. Shoot me an email with your flight information. Can't wait to catch up—

L—

Part VI

Good Things I Wish You

Romance is a peculiar thing. I am renewing old considerations. It changes people so much, often to their disadvantage. When they . . . believe that this is the main thing, and why the world actually exists, then I just can't take it . . .

—Brahms, in a letter to Clara, 1854*

Passions don't naturally belong to human beings. They are always exceptions or extremes. . . . The beautiful and true human being is calm during happiness and calm in times of pain and hurt. Passions must pass soon or one must banish them.

—Brahms, in a letter to Clara, 1857†

---

* Litzmann, *Briefe*, 125.

† Ibid., 205.

34.

THE HOUSE ON INGELSTRASSE was set close to the street, with an arched entryway—sized for a carriage—leading into a central courtyard. There Hart sat waiting for me on a small green bench. As soon as he saw me, he jumped to his feet, pulled me into his arms. The look on his face was open, even joyful. The courtyard functioned as the school's playground, and children raced around us in packs, darting from the sandbox to the jungle gym, from the jungle gym to the swings. There we stood in the midst of it all, staring at each other like two people who had fallen in love.

"I thought you weren't coming," Hart said. "You weren't answering your phone. Did you find the apartment okay?"

"I overslept. Are we late?"

"Not quite."

He led me back toward the archway, where we followed the curved stone stairway to the second-floor rooms occupied by Clara and Robert during the early years of their marriage. Now these rooms held scores and letters, portraits, antique instruments, pieces of furniture pushed out of the way to make room for a dozen rows of folding chairs facing a modern grand piano. A woman was already speaking to the audience, thanking sponsors, mentioning upcoming events, and then Hart lifted his chin as a tall, dark-eyed girl, violin in hand, emerged from behind a curtain.

She wore combat boots beneath a short leather skirt. White sleeveless blouse. A clutter of rings. Short blond hair streaked with chartreuse. Interrupting the crescendo of applause, she planted her feet and attacked the Ciaccona from Bach's Partita No. 2, silencing us, seducing us, working the bright acoustics of the hardwood floors, the plaster walls, to create a sound almost electronic in its overtones. Hart leaned forward in his seat, hands clasped, as if in prayer. Here, then, lived all his lost passion, reverberating around us. Reborn, reinterpreted. Timeless as gold.

At the end of the piece, Friederike bowed perfunctorily, tuned through the applause, and then launched into Kreisler's Praeludium and Allegro without seeming to care that her newly arrived accompanist wasn't quite settled, that the audience was still recovering from the Bach, clearing throats of emotion, shuffling and shifting stiff limbs. No matter. By the third selection, we did not expect her to wait for us, to look at us, to acknowledge us in any way beyond that quick bob of a bow. Only after she'd completed her last number did she glance at the feet of those in the first row, run a ringed hand through her colorful hair, and offer—more to herself than the rest of us—a small, pleased smile. There'd be no encores once she left the makeshift stage, despite hopeful applause that sustained itself for at least three minutes in that small, echoing room.

Already, she'd forgotten us, her violin safely tucked away in its case.

She was back behind the curtain, waiting for the last of the footsteps to retreat down the curved marble stairs. Waiting for the scrape and slap of the folding chairs: picked up, collapsed, put away. Waiting for the accompanist—a local boy, about her age—to enter his number into her cell before leaving with his father who, too, must shake her hand.

Anything we can do for you, they said.

Of course, she said.

But she wasn't really paying attention. She was watching the gap in the curtain. She was freshening her lipstick from a slender tube, running her hands over the creases in her skirt.

Listening for a single set of footsteps.

Thousands of miles away, Heidi had just finished eating Cal's special pancakes for lunch, and now she was sitting on his lap, teasing him, tickling him, begging him for ice cream. He'd take her for three scoops—not Mommy's prescriptive one—and after that, a swim in the lake, without making her sit out the old wives' cautionary hour. That night, he'd keep her up too late, watching movies with her older cousins, roughhousing with her, riling her up, encouraging her wildest exuberance as only a father can.

No, no, she didn't want to go with Daddy. And then when she was with Daddy, she wanted only to stay.

"Are you ready to meet my daughter?" Hart said, taking my hand as if I were the one in need of consolation.

"We're the lucky ones," L— had said as we waited in line at the Starbucks outside the security gate. He looked older than when I'd last seen him, heavier along the jawline, gray hair poking through his shirt collar. Then again, he *was* older. Five years. I was older, too. "No matter what happens in our personal lives, we can always turn to our work."

"But what if that work starts attaching itself to someone outside the work? Or something," I added quickly. "Psychologically, I mean. Let's say I'm inspired by mountains, but I'm forced to move to the Great Plains. Without those mountains—"

"Or without that particular person—?"

"Without that external inspiration, the passion for the work just drains away."

"But isn't inspiration just another word for infatuation?" L— shrugged. "It comes and then it goes. It can't be sustained. You see this all the time with student writing, don't you?"

We placed our coffee orders, joined the other addicts in the close, noisy space beside the pickup counter. The security line was growing. It was only half an hour until my connecting flight would start to board, and I felt that I had so much to say, too much to say that I could say to no one else. "But what is the difference, really," I blurted, "between infatuation and love? I'll be forty-three in another two months. I think I still don't know."

"You're asking me?" L— shook his head. "The guy with the second marriage that lasted, oh, what time is it again?"

He pantomimed looking at a watch.

"Stop it," I said, laughing. "Besides, shouldn't that make you the expert?"

"Point taken," he said, and then he looked at me with such unabashed affection I came to him like a child. "Jeanie, don't you know?" he said. "Infatuation is the inciting incident. Maybe it goes somewhere, maybe it doesn't, but you can't have a story without it. Love is the story itself, the thing we carry with us after the mountains are gone."

I thought about how easy it would be to stay here, in the arms of this good, kind man, who would take care of me, always. But another twelve hours, and I'd arrive in Leipzig. Another sixteen hours, and Hart would rise from a small green bench in a courtyard, kiss me at the entrance to a room where, after her marriage, Clara wrote with such joy: *We love each other more every day and live only for each other.*[*]

* Reich, *Clara Schumann,* 108.

View from the Train, 2006

35.

It's been over a year since their last vacation. It's been two weeks since Robert's death. This time they are traveling to Gersau, unofficially chaperoned by Johannes's sister, Elise, accompanied by Ludwig and Ferdinand, assisted by Clara's maid. They make several stops along the way, including Bonn, where they visit Robert's grave. Clara collects flowers and leaves as Johannes waits for her, openly weeping. When he bends to press his forehead—lightly, sweetly—against her own, it seems to her that she, too, must die: out of sorrow, out of joy, she cannot tell which. For two years, they've been living with the same shared grief, like a lock on a door that cannot be picked. Now there is relief in Robert's death. Both of them feel it. It can't be helped.

Back on the train, continuing south, Clara watches Johannes as he stares out the window, half listening—as she herself half listens—to Elise, who warbles on and on, like a bird. There are times when his beauty catches in her throat like one of his best melodies. There are times when the very thought of his goodness—to her, to the children, to poor Robert—moves her to tears. But then moods strike, like the one which grips him now, and he pulls away from her, avoids her gaze, holds himself aloof. Ever since they crossed the border into Switzerland, he's acted as if she's done something to displease him, though she cannot think what that would be. Or perhaps one of the children said something? For his ill-tempered silence has spilled over onto Ludwig and Ferdinand, who have chosen to sit in second class with the maid, instead of clamoring, as they usually do, to cuddle on the lap of Herr Brahms.

Yet he loves the children, she knows this. Loves them as he loves her. Wants to be a part of them.

A part of her.

Claimed.

So why, then, this terrible coldness? Why does he sit like a stone? All it would take is a word of explanation, but when she ventures to touch his sleeve, he flinches, as if he has never looked upon her with such hunger it took her breath. His mouth open, urgent, against her closed lips. His teeth at her neck, her throat. Month after month, she has turned him away, as she must.

As both of them knew that she must.

Now there's no need to wait any longer. Now the locked door between them is gone. She thinks ahead to Gersau: bathing in the lake, wandering the foothills, playing on the piano at the inn, where they'll live as a family, happy and complete. She is no child. He has proved himself a man. It is only a matter of time until they marry. The next time he comes to her, she won't disappoint him. The next time he kisses her, she won't turn him away.

It was inevitable that he should recognize that the destiny which he had to fulfill was irreconcilable with single-minded devotion to a friendship. To recognize this and immediately seek a way out was the natural outcome of his virile nature. That he broke away ruthlessly was perhaps also an inevitable consequence when one takes his inherent qualities and the nature of the situation into account. But without a doubt he had had a struggle with himself before he steered his craft in a fresh direction, and he never got over the self-reproach of having wounded my mother's feelings at the time, and felt that this could never be undone.

—Eugenie Schumann*

* Schumann, *The Schumanns,* 154.

36.

In the morning, while Friederike practiced, Hart took me to see the house of Mendelssohn, the new Gewandhaus, and, finally, the church in Leipzig where Clara and Robert were married. This was a surprise Friederike had arranged. The caretaker—it turned out—was the father of last night's accompanist, so I was permitted to play the organ, to wander the church as I pleased, while Hart slipped out into the garden, cell phone pressed to his ear. I concentrated on the caretaker's German: a single bronze bell, he explained, still hung in the tower from before the war. Originally there had been three, so when the church was reconstructed, the two stolen bells were replaced. In the end, however, they were taken down, having been forged of iron.

"The sound," he said sadly, "wouldn't mix."

At noon, we met Friederike at her hotel. She embraced Hart and then, to my surprise, gave me a quick hug as well. Had we enjoyed ourselves? she asked as we walked up the street for lunch. Was I able to see what I'd wanted to see?

To me, she spoke in English with a soft French accent.

To Hart, she spoke German or French.

I'd thought she'd be hard-edged, indifferent, but offstage, her shyness was palpable, despite the studded jeans and ripped tank top, hair spiked into a fierce crown of thorns. After the concert, at the first sight of her father, she'd taken three bounding steps into his arms but then, at the last moment, twisted away so I'd found myself grasping her outstretched hand. Hart caught her around the shoulders and neck, the three of us connected in an awkward tableau as, at the same moment, the lights went out. The building was closing for the night.

"Attendez!" Friederike called.

"Pst, vielleicht können wir hier schlafen!"

And Friederike, not knowing if I'd understood, said, "Papa thinks we should spend the night. What a marvelous thing for your book!"

In the end, we flew, hollering, down the stairs, surprising the old docent with his ring of keys. Hart held Friederike's violin aloft. He was bright-eyed, reckless. "Catch!" he said, pretending to toss it, after we'd made it outside.

"Papa, you are crazy."

"Here it comes!"

Now Friederike told me that her violin was insured for one million euros. "Papa bought it for me two Christmases ago," she said, glancing out the window at the street, where Hart had gone to address another round of calls from Lauren. Meanwhile, we lingered over coffee and dessert, the first time we'd been alone.

"You're lucky," I said, trying to align a gift of this magnitude with so many years of silence. "Few people understand how important it is to have a really good instrument."

"He set it up to look as if it came from a group of donors, but Maman's attorney looked into things and figured out that Papa was—relating to it?"

"Behind it."

"*Oui*. At first Maman wouldn't let me accept, but really, what could she do? Since then, she is always checking up on me, worrying, opening my mail. Of course she is angry about things, still. But he's told you everything, I am sure. And it happened so long ago."

"He has told me," I said, choosing each word with care, "that both he and your mother wanted very much to raise you."

"But does anyone ask what *I* want?" She said something in French that I couldn't understand. "I want to go to New York City, to study at the Juilliard. Maman is getting married again. She will not be alone." She nodded so I'd know Hart was on his way back to the table. "Perhaps you and Papa will marry as well. He tells me of your little daughter, too. A wonderful pianist, very clever."

What I thought: *Whom he met only once.*

What I said: "Very clever, when she practices."

"What have you been talking about to please our Friederike so?" Hart said, sliding into his seat beside me, covering my hand with his own.

Our Friederike.

How it thrilled me, how it frightened me, to belong once again to this tableau: the mother, the father, the beautiful child. People smiling to us as they passed.

We rented a car, drove to Zwickau. We spent several hours at the Robert Schumann House, then had supper at a small, ghastly inn where at one point, for no discernible reason, an enormous painting fell from the wall in an explosion of glass and dust. Moments earlier, Hart had been mocking it; now, he and Friederike could not stop laughing. When they saw me hunched beneath the tabletop, they laughed even harder, Friederike wiping mascara from the corners of her eyes.

"Pourquoi est-ce qu'elle se cache comme ça?"

"She is American," Hart said as waiters rushed in, exclaiming over the mess, "and all of these crazy Americans, you see, are carrying guns wherever they go. You hear a sound like that in Miami? Everybody ducks."

"It's *not* that bad," I protested. "And we do *not* all carry guns."

"You want to fuck wit' me?" Hart said, doing his best Scarface, which wasn't quite as good as he thought. "*You* want to fuck wit' *me*?"

"Papa!"

Even before we were asked to leave, Hart was throwing euros on the table, and we stumbled out into the cobblestone streets, drunk on wine and bad behavior. Our epic quest: another inn. We wound up driving through the countryside. Hours later, in a little town balanced at the far edge of the earth, I pointed out what appeared to be a very nice resort. Friederike was trying to tell me something, but Hart pulled into the long, circular driveway, beautifully manicured with sculpted trees, suggesting I see about rooms. I got out, tried the arched front door. Locked. A woman dressed in white passed through the well-lit halls. I banged on the door to get her attention. She stopped, froze, looked at me, then sternly shook her head.

"Bitte?" I called through the glass. *"Hotel?"*

It was at this point that I heard Hart laughing.

"Papa, you are terrible!" Friederike said.

I was standing in front of a mental institution.

This, then, is the story I'll take with me. Not the museums in Leipzig. Not the carefully collected curios of Zwickau. Not the Endenich asylum, the elaborate grave in Bonn, the house in Düsseldorf with its circuitous route to the Rhine.

Finding, at last, a place to sleep. Friederike's shower running in the room next door. Hart watching me from the narrow bed. Making love well beyond the point of exhaustion when, still, there is somehow more to be given, more to be received.

If only I could tell you, now, all the good things I wish you. Long out of sight of the mountains, what I still carry with me.

This.

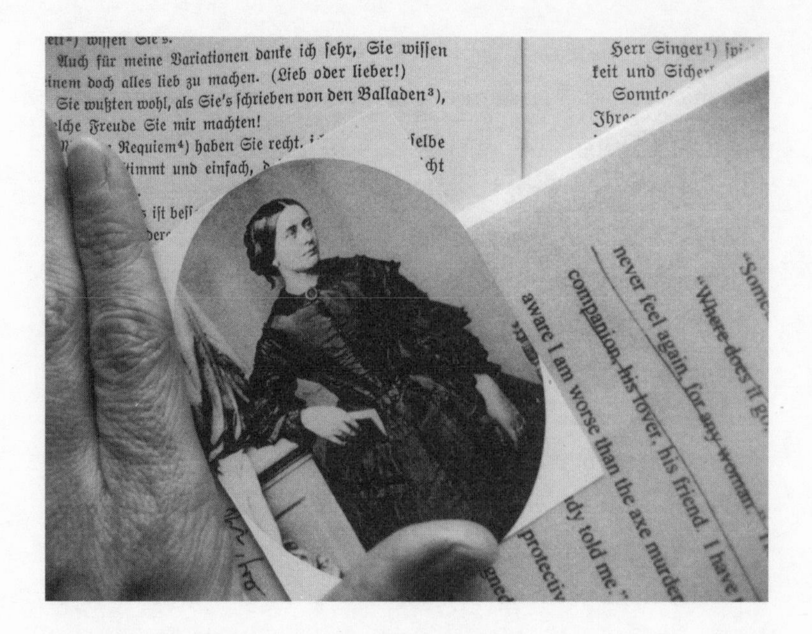

I suffer indescribably at being separated from Johannes . . . And yet is it not most natural that I should love and esteem Johannes so much, after such a long and intimate relationship with him, during which time I learned to know fully the riches of his heart and mind?

—Clara, in a letter to Joseph Joachim, 1857*

---

* Swafford, *Johannes Brahms*, 164.

37.

We were settled in our Zurich hotel by the time my cell phone rang. Friederike's music camp had started just that morning; I was writing at the marble-topped kitchen counter while Hart made phone calls from the desk in our bedroom, switching from English to French and back, speaking with his American attorney, his French attorney, Lauren. I dug the phone out of my purse without looking, eager, expecting Heidi's clear voice. "Sweetie, how are you?" I trilled.

"*You* sound awfully cheery."

"Ellen?" I said, disoriented.

"Getting lots of research done?"

I looked around the beautiful suite: curved couches and monogrammed robes, a small grand piano, balcony doors opening onto a wide view. "Yes and no."

"Well, I've been hard at work. Would you like to hear what I've found out about your German doctor?"

Something deep in my chest gave way; I took the phone into the bathroom. "Okay," I said, sitting on the edge of the oversize whirlpool tub.

"He got his medical degree in East Berlin, but afterward he stayed on to do research. After the wall came down in '89, he left his university and formed Viso-Tech."

Our connection was poor; I struggled to hear over the crackles. "He told me all this—" I began.

"—ever Google this guy? He was one of those corporate cowboys everybody loved in the nineties. Profiles in *Forbes, CEO, People.* He came up with this implant that goes into the brain, helps it process signals from the eye."

"Someone was telling me about it."

"Anyway, he gets only good press until August of 2000, when he's arrested at Disneyland."

"Where? I'm having trouble hearing—"

"*Disneyland.* For kidnapping his daughter."

"*What?*"

"Six years ago, he went to France, picked her up for the weekend, then never took her back to her mother."

"Are you sure this isn't exaggerated? You know how these accusations get tossed around during a bad divorce." But was I thinking about Hart's limited visitation, about Lauren's reaction to the gift of the violin.

"Hello? You're cutting out."

"It's just—we've been with Friederike all week, and she's made it pretty clear she wants to live with her dad—"

"If Cal disappeared with Heidi, would you be making excuses like this?"

Fat chance. I sighed. "Anything else?"

"He's still on the Viso-Tech board of directors, but his involvement is minimal these days. He's listed on staff at

Bascomb Palmer. He does a little teaching, a little public speaking."

"So, in other words, he's basically who he says he is."

"Aside from the kidnapping charges, yes."

I didn't say anything.

"Oh, there's one more thing. A few years ago, he was hospitalized in Santa Fe. Another little bubble of publicity. Since then, there isn't much written about him. He's published a few papers. Lectured at a few conferences."

"What was he hospitalized for?"

"Maybe you should ask him."

"Or maybe it's none of my business."

"You're angry at me."

"You're angry at him. I'm trying to figure out why."

She must have moved because suddenly the connection was perfectly clear. "He tells you not to get attached to him, but he introduces you to his daughter as if you're someone special. He says there's no chemistry between you, but, let me guess, you're in bed all the time. And I know exactly what you're thinking. You're thinking *he's* something special, this tragic, troubled man. You're thinking that, at some point, he'll recognize everything you are to him and love you in return. Let me tell you from experience: both of you can't be in love with him. A man like that cares about no one but himself."

38.

On the train to Lucerne. Late afternoon. Hart sat beside me in the window seat, staring out at the dark, textured surfaces of the mountains, at the sky which was still the bright blue of morning. Tomorrow, after our night in Gersau, we would visit the glider club in St. Gallen. Everywhere we went, he watched the scatterings of clouds, assessing conditions. Would there be lift? I knew exactly what he was thinking, and yet I felt no closer to knowing this man than when we'd met three months ago. The ax murderer. The kidnapper. Three marriages.

You mustn't get attached.

Suddenly I was missing Heidi terribly. I checked my cell, but there was still no message, no text, no photo. What would I do if I returned from Europe to learn Cal had taken her out of the country? Started a new life with her in Vancouver, Panama City, Tibet?

"Who were you speaking to this morning?" Hart said.

"A friend of mine from the States."

He was watching me with the same expression with

which he'd been studying the sky. "Whatever he said upset you, I think."

"She."

"Ah."

"She ran a background check on you."

"At your request?"

"Her own initiative, actually."

"An enterprising woman."

"More like a concerned friend."

"And what, in particular, concerns her?"

What I thought: What doesn't?

What I said: "You tell me."

Behind his shoulder, the countryside flashed by: open fields populated by cattle and goats; steep vineyards; exposed rock; scattered outcroppings of slope-shouldered pine. Beautiful countryside. Beautiful land. I wanted to scoop it up by the bucketful, drink it down in gulps, wash away my suburban South Florida life of billboards and strip malls, asphalt and road trash, terrible traffic and the sort of heat that begins to feel something like rage. Even here, it clung to me still, like burned and blistered skin. I wanted this place to be the one I called home, even as I struggled to speak its language, even as I knew that, in another week, I'd be back in the world where, no doubt, I'd come to belong. Neither vibrant city nor lyrical country but the undefined space in between. Staring out my window at the artificial lake. Walking the beach with my face turned toward the water, away from the condos, the shops and parking lots.

"What is there to tell?" Hart said. "I should not have taken her out of France. I should not have kept her so long."

"Where did you take her?"

"La Scala. The London Philharmonic. Tanglewood." He cocked his head. "What, you were expecting a dungeon?"

"I wasn't expecting *Disneyland*."

"It was where she wanted to go. Our last stop."

"Your last stop before what?"

"Before it was time to take her back home."

"You were going to take her back to France?"

"I'd sent Lauren our itinerary. She turned it over to the French authorities, who turned it over to the Americans. That is how I came to be arrested in front of—what is that ride? The one with the planets? Space station?"

"Space Mountain?"

"Yes, it is just this otherworldly. They put me into handcuffs. You can imagine the mess. I lost interest in my research, I still cannot explain. I took a leave of absence. I traveled out west for a year."

"Where you ended up in the hospital."

"Your concerned friend, I'm sure, has told you about that, too."

"Not really," I said, but he was getting to his feet.

"Quick, before I kidnap you. We are coming to our stop."

The last ferry was about to leave; we bought our tickets, stepped aboard. Minutes later, we were sliding through the green waters of Lake Lucerne. Mount Pilatus behind us. The Rigi like a muscular shadow ahead. Upstairs in the dining room, Hart surprised me by ordering champagne.

"Now, for happier topics," he said as the waitress put out our glasses. "Friederike attends the Juilliard this fall. It is settled to everyone's satisfaction. So it's doubtful, I think, that Swiss guards shall arrive—tonight, at least—to carry me away."

"Congratulations," I said.

"Yes, yes, it is a wonderful thing. I will relocate to New York, buy an apartment, pay a housekeeper to greet her after school and care for her when I travel. Of course I will cover Friederike's tuition and expenses. Meanwhile, should Lauren feel the emotional loss of her daughter too intensely, I shall continue paying child support as a means of cushioning such strong, maternal grief. Should her soon-to-be husband benefit from such considerations, well, such an effect must be considered incidental."

"You will move to New York," I repeated. For some reason, this hadn't occurred to me: that Hart would be going with her.

"In exchange, Lauren does not sue me in court for more alimony, to which she is likely entitled. My attorneys assure me it comes out the same. Ah, the sweet fruits of justice."

"You have plenty of money," I said quietly.

Hart looked at me. "This is true."

"And now you have your daughter. So, in the end, you are the lucky one, just as you always tell me. You are the one who has everything."

"Not everything, no." He refilled his glass. "But perhaps you would agree to come with us? You told me once that you missed living there."

"And my suitcase is already packed. So why not?"

"Your daughter will come too, of course. There you'll have access to the sort of teachers, the sort of environment . . . well, isn't that what you are wanting for her?"

I stared at him. "You are serious."

"Come, how difficult can this be? I will get to know your Heidi, she will get to know me." He topped off my champagne. "As for your teaching, I am guessing you could find another position, if you wish. But wouldn't you rather just write your books? Did I ever tell you the story of how Friederike came to play the violin?"

I shook my head, still trying to wrap my mind around the question at hand.

"She was eight. Lauren and I were just married for the second time. A colleague of mine invited us for a week in the Bordeaux countryside. One night at dinner, someone starts to play the piano, and we're thinking it must be his daughter when the daughter herself walks in. *Friederike won't take turns,* she announces, and I get up to see my Friederike playing the same piece this girl had performed for us just the night before. All the way back to Paris, Lauren and I argue about what to do. *Anything but the piano!* she says. *It will only spoil her hands.* In the end, Friederike gets a violin."

"Well, it seems to have worked out."

"In New York, you could supervise Friederike's practicing. You could be there to care for her when I must be away. You could advise her as only a woman can."

I pushed back my glass, dismayed. "Are you asking me there as your lover or as the household manager you're reluctant to hire?"

"Neither. Both." He looked at me earnestly. "I am asking you there as my wife."

"Are you absolutely crazy?"

"A reaction I've not yet experienced," he said drily.

"We've known each other three months!"

"How long did you know Cal before you married him?"

"Three years."

"I knew Lauren two years the first time, ten the second. For all the damn good it did us. Come, we both have attorneys. We will write a solid prenup. I get along with you, I suppose, as well as I've gotten along with any woman."

I flushed. "How romantic of you to say so."

"Don't be that way, Jeanette," he said, and I saw he was hurt, too. "Would you rather I tell you a lie? What can romance mean at our age? I am thinking we understand each other."

The town of Gersau appeared on the horizon: an onion-headed steeple, surrounded by houses set into the mountainside. At a restaurant across the street from the station, we asked for directions to the town's only inn. By now the light was fading fast, and together we walked up the steeply sloped path as he talked of music and sunlight and trees. Of stones, of sour cherries. Of history and God, diseases of the nerves, augmented chords, a restaurant in Dresden where, once, he had sampled a particular kind of cheese. But I couldn't keep up with all that he said. I couldn't keep up with my own flooding heart. This was not the sort of man with whom one builds a future. And yet I could not step away.

"Why were you hospitalized?" I asked.

"We are back to that."

"We are back to that."

Around us, the darkening land. The fading light of dreams.

"Long before reading poor Schumann's story, I had the misfortune to jump off a bridge."

"Stupid," I said. It just came out.

"Yes, my dear, strong Jeanie," he said. "Stupid indeed. Particularly when you consider how very well I swim."

Gersau, 2006

39.

"IT IS SOMETHING OUT of my childhood," Hart said, staring at the narrow bed, the washbasin, the window shutters thrown open in the hope of attracting the slightest breeze. Cooking smells thickened the air. The bath was down the hall. The overhead light didn't work, but there was an electric candle on the nightstand and—another touch of whimsy—a small golden cherub mounted above the bed. "Is this the inn where they stayed?"

"Possibly. The house was here, but I don't know which years it served as an inn."

We took turns washing up for the night, then lay naked on top of the sheets. Too hot to touch. Too hot to sleep.

"I wish," Hart said, "I had something to read."

"I've got a book, if you want it," I said, reaching for the short-story anthology I still hadn't managed to finish.

"You do have a book, yes." Hart took the anthology, placed it like a brick between us. "But I'm thinking of the one on your laptop. When do I get to read what you are writing?"

Our bodies looked strange in the false candlelight, each of us neither old nor young. "I should tell you something first," I said. "Something I should have told you before."

Hart touched the anthology with one finger. "I've never enjoyed any conversation which begins this way."

"I've been writing about you," I said.

For once, he had nothing to say.

"Not *you*, of course," I added. "Everything changes when it hits the page. But there are things you're going to recognize. Not because they're true, but because they're not. The same way that anyone who knows Clara's life will recognize—"

"Clara's life, yes. A woman born in 1819. I thought you were writing a historical novel."

"I thought so too."

"What happened?"

What I thought: *I was a dead woman. I was a stone.*

What I said: "There are things between men and women that do not change."

×

They will spend this day as they've spent all the others: hiking during the relative coolness of the morning, bathing away the afternoon heat, then a few hours at the piano, taking advantage of the lingering light, taking advantage of a language in which everything between them is always understood. Sound floating over the lake like mist, climbing the steep, crooked paths between houses where conversations still to listen, where overtired children release themselves to sleep. Somewhere a nightingale repeats its circuitous lullaby. Higher up, in rough-cut pastures, drowsy-eyed cattle add the tinkling of bells.

How quiet the evenings in the absence of carriages.

How silent the nights with their showerings of stars.

Inside, as always, the air is motionless, stifling. Clara removes her dress and stockings, splashes her face with lake water. She hears Elise across the hall, still talking as the maid turns down their beds. Ferdinand and Ludwig are already sleeping in the room they share with Johannes, who makes his own soft music as he washes, moving about, humming.

It's been a good day for them all. At last Elise is silent.

A window shutter creaks.

There comes the good smell of Johannes's cigar, and she leans out her window, too. No need for candles; the moon is full, fierce.

They smile at each other in the clarity of its light.

Usually, they walk onto the balcony, settle themselves in rocking chairs that face the water and, beyond the water, the foothills of the Alps. Johannes carries his candle for them both, a single, flickering heart, and at some point, they find themselves holding hands. For the rest of her life, she will think of that hand, small and smooth and remarkably delicate. A woman's hand, if she did not know better, dwarfed by her own massive palm. She will think of the smell of Johannes's cigar, the high, boyish murmur of his voice. The faint lapping sounds of other voices, other balconies, other lovers waiting for the heat to ease, returning to their beds, turning to each other, houses so close that this music, too, becomes part of the night, part of all of the nights in which Clara has lain sleepless, aching, a single thin wall the only thing that divides her from what she wants.

Tonight, she is tired of waiting. Tired of longing. This terrible restlessness that won't let her sleep. The thoughts that torment her, unspeakable, consuming. Perhaps she is tasting what poor Robert tasted. Perhaps she too is going mad. Certainly it is madness that prompts her to take Johannes by the hand and lead him not to the balcony chairs, but down the wooden steps into the landscaped courtyard, through the gate and onto the stony path they've climbed so many times with the children in the freshness of morning, in the brightness of day, gathering flowers, admiring the scattered outcroppings of rock embedded in the Rigi's green face. Perhaps it is the wide, curious eye of the moon that propels her to walk faster, even as she feels him start to resist, as the silence between them grows shadowed and thick, uncomfortable as the humidity, oppressive as the heat.

Even now, this heat does not break.

Another gate. Cows like pale boulders.

A fence. The first stand of trees.

More trees, and now woods, moon dappled, fragrant, where she guides him toward a platform she and the boys found yesterday, placing their feet in the ax-hewn steps, hauling themselves up into the cooling leaves. For a moment it seems they can both believe this discovery is why she has led them here. Johannes is delighted. He shimmies up the tree trunk to the platform, then continues to climb, working his way from branch to branch, his dressing gown sending showers of snagged twigs and leaves upon Clara's head. Even after stopping to knot her gown, she must struggle to climb each step. She is a woman. She is no longer young. When she reaches the platform, she sprawls on her back.

"Johannes?" she calls, but he does not come down.

×

Hart was reading, the way he did read: silently, rapidly, utterly absorbed. I held the anthology open on my stomach. I watched the cherub, who watched me back.

×

"Johannes?"

She holds her breath, listening. Though she can't hear him, can't see him, she knows he is just above, listening too. Suddenly, wildly, he drops onto the platform; she gasps as the boards shake beneath them, and then he is laughing, she is scolding, they are both themselves again. So much themselves that it seems, at first, that their hands on each other, their mouths on each other, are just a continuation of another conversation, more of the music they make together so easily, so effortlessly, only now everything between them turns strange, his breath like a curse, a hiss, a howl, and it's nothing like what she remembers of tenderness, over almost before it's begun.

His gaze on her is a stranger's gaze: terrible and cold.

"What is it? What is the matter?" she says.

He thrusts her away with such force that she nearly tumbles from the platform's edge.

She leaves him there, stumbling down through the trees, across the pasture and along the rocky path toward the garden gate, which she passes, and the church, which she passes, and the little road fronting the water, which she crosses, and then she is standing at the edge of the lake. Up to her ankles. Up to her knees. The moon lighting up her shimmering reflection. Angular body. Strong-featured face. She thinks, as she's often thought before, that if only she'd been prettier, more feminine, everything would have been different. If only she'd been less gifted, less determined, less strong. What happened just now serves only to confirm this.

What happened to Robert was her fault.

×

"Was there really a tree house and a platform?" Hart asked. The clock in the corner of my laptop screen said

2 A.M. "Did she really walk out into the water that way?"

"I made up everything in that scene."

"The way you made things up about us."

"But it *isn't* us. That's what I'm trying to say."

"Isn't it?" He rubbed his eyes, and when he looked at me again, I saw the white-coated scientist in his lab, the man I'd first met at the Wine Cellar. "The scene is effective," he said, "because it plays a trick on the reader. We are not expecting she'll be thinking of her husband. I myself had not thought about this before. But I believe you are right about how she felt."

That heat: pressing down like a smothering hand.

"It is the same way, I suppose," Hart said, still watching me closely, "you feel about your Cal."

Hard to get breath. Hard to speak. I was about to say it was the last thing I'd been thinking of when I realized it had been the only thing.

All along.

Even now.

"So you see," Hart said, "you did not make everything up after all."

×

Slowly she climbs the path to the inn, her wet gown chafing, sticking to her legs. He is waiting for her on the balcony: smooth faced, slim shouldered. He, too, has been crying. She sits beside him; he extends his hand. She takes it—what else can she do? At last they understand each other. Or perhaps she has understood him all along. He cannot love her, love anyone, completely.

This time, she is protected. Safe.

The first, cool fingers of breeze stir the air. The horizon goes gray, then pink. Clara feels something loosen around her heart, the first flaking pieces of a grief even older than Robert's illness lifting away. Come September, she'll be back on the road. She will travel to England, to Russia again. She will, she believes, see America this time.

She will, she believes, feel whole and well again.

×

When Hart turned out the electric candle, the whole world vanished in darkness. The sort of absolute darkness in which anything might be said. "You don't love me," I said. "You have tried, I know. But you can't."

"There are better things, more important things, than the kind of love you are meaning." He turned in the bed to face me; his breath was strangely sweet. "Here is the truth, which I've tried to tell you ever since the second time we met. I don't think I'll ever love anyone the way I loved Lauren. I'm not sure now I would want to. One loses too much to a love like that."

I thought of how Clara's music suffered during the happiest years of her marriage. How Robert protected his art from Clara's love behind the closed door of his studio, behind the closed mind of his madness. The long years of sadness between Cal and me, during which time I published five books in six years.

Yes, Hart understood me. I understood him, too.

Art is about desire, is it not?

I'd chosen him, exactly, for this.

Date: Sunday, July 23 2:52 PM

To: Jeanie88@comster.com

Hi, Jeanie—

There's this couple in their sixties and they've just gotten married and they decide that they want to have children. So they go to a fertility doctor, and he's doubtful, but hey—cash is cash—so he hands them a container and says, Let's get a sperm sample first. One hour later, the doctor comes back into the room, only to find the container is still empty.

"What seems to be the problem?" he says.

"Well," the man says, "first I tried with my right hand, and then I tried with my left hand, and then my wife tried with both hands . . . but we just couldn't get the cover off the container."

In the end, I suppose, it all comes down to this.

Much love to you always, wherever life takes us.

L—

40.

BREAKFAST AT THE INN was coffee and bread, fruit and cold meat, cheese. Hart was already on his way to St. Gallen. I'd walked him down to the ferry dock, where we'd kissed each other good-bye. But I hadn't wanted to spend the day flying with him, and he, in turn, hadn't pressed. Something between us was already different. Some necessary tension had eased. At the end of the week, we'd meet in Zurich, in time for Friederike's recital. Back in the States, we'd talk on the phone. I'd see him until he moved to New York. Perhaps, now and then, after that. And yet, as I'd walked back from the ferry dock, I felt as if I'd just been to a funeral. I saw him as if he were still beside me. I could hear his voice in my head. The room smelled, lightly, of his cologne. The shape of our bodies still hugged the bed.

After breakfast, I reserved the room for a second night, and then I wandered along the few waterfront shops until I found a bench overlooking a strip of rocky beach. My eyes blinked, burned against the fresh morning light. Déjà vu. At last I understood what the feeling meant, what it was I had recognized. It hadn't been Hart after all. It had been what he'd made me feel. Alive to the world around me. Alive, once again, to myself.

My cell phone buzzed. Heidi, at last. I cleared my throat, answered. But no—it was just Cal.

"I didn't want you to worry," he said. "She just doesn't want to call you this time. I'll try to put her on, though, if you like."

"No, it's okay. Is she happy?"

"Very happy. Settled."

"If she's happy, I'm happy."

"You don't sound happy. You sound kind of awful."

How I wanted, at that moment, to tell him everything, to let him comfort me, console me, the way he'd done countless times during our married life together. All that I'd lost truly hit me then. I was alone in the world, I was truly alone. And yet, I must raise Heidi. I must go to work and come home. I must shop and cook and clean the house, balance the checkbook, take the car in for oil changes, repair the gutters, pay taxes. A privileged life, a blessed life, in a world filled with hunger and terror and want. It was shameful to admit, even to myself, that it all seemed impossible somehow.

"I've picked up a summer cold," I said.

"That's too bad."

"It'll pass," I said. "Thanks for calling."

"Listen," Cal said. "I just wanted to say I'm sorry. For the way I acted the last time I saw you. It isn't even true, what I said."

"What you said?" I was trying to remember.

"Of course my girlfriend reminds me of you, in some ways. I mean, you and I had so many good things between us. Good years. We still do, don't you think? I mean, after we get through this part."

"After we stop being mad at each other," I said.

"I want to stop."

"I want to stop, too."

"God, you really sound terrible."

"I better hang up, I'm losing my voice. Give Heidi a kiss for me?"

It seemed as if I'd never look forward to anything again. Upstairs at the inn, I pulled the sheets over the bed, closed the thin curtains against the heat that was already spreading over the walls, thickening the moist, still air. I wanted so much to lie down in that bed, close my eyes and bury myself, disappear for the day. Instead I sat down in front of my laptop and turned to the work at hand, to the writing that has always sustained me, kept me whole, even when everything else around me falls apart.

"My true old friend, the piano, must help me," Clara wrote in 1854, shortly after Robert was institutionalized. "I always believed I knew what a splendid thing it is to be an artist, but only now, for the first time, do I really understand how all my pain . . . can be relieved only by divine music so that I often feel quite well again."*

---

* Swafford, *Johannes Brahms*, 119.

Perhaps, in the end, it was this belief that formed the cornerstone of their friendship. Clara Schumann could not have been the pianist she was, nor Johannes Brahms the composer he would become, had they not shared the same need to remedy longing.

To medicate loneliness.

41.

Brahms on the train back to Hamburg, then. Still young, still slender, but already beginning to carry himself with the reserve of a much older man. Hurtling toward his next love affair, and the love affair after that, toward the series of women he will choose because each is impossible to attain. Running through the maze of his heart's desire, one which will lead him, again and again, to the same plain-faced, plainspoken mother of seven children, fifteen years his senior, mirror of his genius. Home. He will dream of her worn face, the fullness of her body, the broad weight of her hands. He will long for her even as he hates himself precisely for that longing. He will refuse himself the consolation she offers, as well as the consolation of her children, burdened not so much by disappointment as the constant expectation that things will go wrong.

Women are fickle. Whores grow annoying. The young pretty girl who attracts him has nothing of interest to say.

Years later, upon learning she is dying, he'll clutch at his heart and cry out to a friend, "Apart from Frau Schumann, I am not attached to anyone with my whole soul! And truly that is terrible and one should neither think such a thing or say it! Is that not a lonely life . . ."*

"Although he loved human intercourse and sought it," Eugenie Schumann would recall, "he was always on the defensive when he was sought after. He liked to give, but resented demands or expectations. He selected his friends very carefully, and there were not many who passed muster. Once, during the last years of his life, he went so far as to say, in an outburst of moodiness, 'I have no friends, and if anyone tells you he is my friend, don't believe him!' We were speechless. At last I said, 'But, Herr Brahms, friends are the best gift in the world. Why should you resent them?' He looked at me with wide-open eyes and did not reply. Brahms undoubtedly has suffered very much. He was very human, as human as one can be, and that is why he was also much loved."†

---

* Swafford, *Johannes Brahms*, 611.
† Schumann, *The Schumanns*, 156.

How much I have possessed and how much lost! Where did I find the strength to live on for so much longer, and to keep working? Whence comes the courage of human beings? The children—and the artist life; it is their love and the compulsion of art that have borne me up . . .

—Clara, in her diary, 1884[*]

---

* Harding, *Concerto*, 226.

The thought of my D-minor sonata proceeding gently and dreamily under your fingers is so beautiful. I actually put it on my desk and gently, deliberately accompanied you through the thickets of the organ point. For me there is no greater pleasure than to be always at your side . . .

—Brahms, in a letter to Clara, 1889*

* Reich, *Clara Schumann*, 197.

Sources

Bickley, Nora, editor and translator. *Letters from and to Joseph Joachim*. London: Macmillan, 1914.

Burk, John. *Clara Schumann: A Romantic Biography*. New York: Random House, 1940.

Harding, Bertita. *Concerto: The Glowing Story of Clara Schumann*. Indianapolis: Bobbs-Merrill, 1961.

Holde, Artur. "Suppressed Passages in the Brahms–Joachim Correspondence Published for the First Time." *The Musical Quarterly* 45, no. 3 (1959): 312–24.

Litzmann, Berthold, editor. *Clara Schumann–Johannes Brahms: Briefe*, vol. 1. Leipzig: Breitkopf und Hartel, 1927.

———. *Clara Schumann: An Artist's Life, Based on Material Found in Diaries and Letters*. Translated by Grace E. Hadlow. 2 vols. New York: Vienna House, 1972.

———. *Letters of Clara Schumann and Johannes Brahms, 1853–1896*, vol. 1. New York: Vienna House, 1973.

Nauhaus, Gerd, editor, and Peter Ostwald, translator. *The Marriage Diaries of Robert and Clara Schumann: From Their Wedding Day Through the Russia Trip*. Boston: Northeastern University Press, 1993.

Reich, Nancy. *Clara Schumann: The Artist and the Woman*. Ithaca and London: Cornell University Press, 1985.

Reich, Susanna. *Clara Schumann: Piano Virtuoso*. New York: Clarion Books, 1999.

Schumann, Eugenie. *The Schumanns and Johannes Brahms: The Memoirs of Eugenie Schumann*. Lawrence, Mass.: Music Book Society, 1991.

Swafford, Jan. *Johannes Brahms: A Biography*. New York: Vintage Books, 1999.

List of Images

Notes and Acknowledgments

All clutter-collages were assembled from published texts (acknowledged under Sources), quotes and translations, translations in progress, journal entries, handwritten notes, props, old portraits, new photos—in other words, everything and everything that cluttered my desk during the writing of *Good Things I Wish You.*

I am grateful to Gaela Erwin for permission to use, on page 70, one of her many exquisite and unique self-portraits. Thanks to Michael R. Ansay for the use of his photograph "Interior" on page 112.

The quotations from Berthold Litzmann's German edition of *Clara Schumann–Johannes Brahms: Briefe* (cited as *Briefe* throughout the text) are original translations by Winfried Reichelt.

I would like to thank my first reader, Sylvia Ansay, as well as Dick Ansay (who else could find a laser printer—on sale!—in the mountains of North Carolina?); Carolyn Broadhead; Jan Conner (for stories and a quiet place to write); Preston Merchant and CJ Hribal (for good advice and kind words); Stewart O'Nan (with whom I wrote, over ten years ago, a screenplay about Clara and Brahms);

Felicitas Reichelt; the Ragdale Foundation; and the faculty, students, and staff of the MFA program at the University of Miami in Coral Gables, especially Lydia Starling and Pat McCarthy. Shari and Tom Goodmann: this book could not have been written without the music of your friendship and good advice.

I'm grateful to Jake Smith for recovering this manuscript after my old computer crashed and setting me up on a new computer (which had been sitting under my desk for a year because I didn't have the energy to deal with it). Jake also developed the artistic design of my Web site. Thank you to Tim and Naoko Soderberg for help with my Japanese English and for suggesting the beautiful name Midori. Thanks to Laura Schalk for helping me out with French. Albert Uster, no longer with us, gave me all the right directions for finding exactly the Gersau I wanted to see.

Thanks to my life mentor, Deborah Schneider, for the best line in this book. Thanks to Claire Wachtel and HarperCollins for letting me take a few chances. Thanks to Sanna Mehlin Tilley, Marissa Matteo, Carla Garcia, Yolanda Jiminez, and, most especially, Ariana Ramieri.

And thank you, Genevieve Ansay, for suggesting the pitch-perfect name for the daughter in this book.